Petroleum Engineering and Development Studies

Volume 2

Directional Drilling

Petroleum Engineering and Development Studies

Volume 2

Directional Drilling

T. A. Inglis

Graham & Trotman

A member of the Kluwer Academic Publishers Group

LONDON/DORDRECHT/BOSTON

First published in 1987 by

Graham & Trotman Limited	Graham & Trotman Inc.
Sterling House	Kluwer Academic Publishers Group
66 Wilton Road	101 Philip Drive
London SW1V 1DE	Assinippi Park
UK	Norwell, MA 02061
	USA

Petroleum engineering and development
studies.
Vol. 2: Directional drilling
1. Petroleum engineering
I. Inglis, T.A.
622'.3382   TN870

ISBN 0 86010 716 7 (Vol. 2, hardback)
ISBN 0 86010 932 1 (Series)

© T. A. Inglis, 1987

LCCCN 87-12110

Typeset by Acorn Bookwork, Salisbury, Wilts
Printed in Great Britain at the Alden Press, Oxford

FOREWORD

Some 35 years ago I was somewhat precariously balanced in a drilling derrick aligning a whipstock into a directional hole in North Holland by the Stokenbury method, and no doubt thinking to myself that I was at the very forefront of technology. During the intervening period it has become obvious to many of us that some of the most significant technical advances in the oil business have been made in drilling, and particularly in the fields of offshore and directional drilling. It has also become apparent that the quality of the technical literature describing these advances has not kept pace with that of the advances themselves in many instances. A particular glaring example of this has been in the field of directional drilling where a large literature gap has existed for many years. I am delighted to see this gap now filled with the present volume by my friend Tom Inglis. Indeed it is only after reading his comprehensive book that I realise the extent of my own ignorance of the latest techniques of directional drilling and how desirable it was to have an authoritative text on the subject. I feel sure that this volume will be welcomed by the industry and warmly recommend it to all who are in any way involved and interested in the fascinating world of drilling.

Jim Brown,
Emeritus Professor of Petroleum Engineering,
Heriot-Watt University,
Edinburgh

PREFACE

To those outside the drilling industry, and even some within it, directional drilling may appear as a black art. It is shrouded in its own special tools and procedures, and even has its own terminology. For the uninitiated it is not easy to come to terms with this mysterious subject. Yet without directional drilling much of the world's offshore oil and gas reserves could not be developed economically. It is fundamental to the production of oil from the North Sea and the Gulf of Mexico, as well as many onshore areas. Anyone who is involved in such developments, be they geologists, design engineers, managers or accountants should be aware of the benefits and limitations of this technique. Outside the oil industry, in such areas as mining and geothermal energy, directional drilling has an important role to play.

Directional drilling has a long and interesting history, stretching over the past hundred years. However, it has only been in the past ten years that much of the new innovative technology has been introduced to meet the challenges of offshore developments. This rapid growth in directional drilling has justified a text-book which gives more prominence to the subject.

This book is intended for students following courses in petroleum, mining or drilling engineering. It will also provide a good introduction for those employed by oil companies and service companies engaged in directional work. As with any important topic it is difficult to encompass all aspects of the subject within the confines of a text-book. I have had to leave out many details in order to keep the book to a reasonable length. I hope that those requiring further information will find it from the references at the end of each chapter. For those readers who do not have a drilling background I have included a glossary of some of the more common terms used. The final chapter is intended to bring the reader up-to-date with the new technology being introduced. The speed with which these ideas are being implemented, however, may mean that by the time these pages are read they could more accurately be described as "current" rather than "future" developments.

This book could not have been written without the help and encouragement of many people. My thanks go to the many service companies who provided literature and illustrations, to the Society of Petroleum Engineers who gave permission to reproduce tables and figures from published papers, to John Thorogood of Britoil and David Knox of Shell who reviewed the text and made many useful comments, and to Professor Jim Brown of Heriot-Watt University who very kindly agreed to write the Foreword.

T. A. Inglis
November, 1987

CONTENTS

Chapter 1

INTRODUCTION

Directional drilling has been described as "the art and science involved in the deflection of a wellbore in a specific direction in order to reach a pre-determined objective below the surface of the Earth." How much art and how much science is actually involved is still debatable, although there has been a significant swing towards scientific methods in recent years.

The origins of directional drilling in the oil industry go back to the late nineteenth century in the United States. Rotary drilling techniques were being introduced, replacing the older cable-tool rigs. At that time little attempt was made to stabilize the drill string and so control the path of the wellbore. Borehole surveys taken some years later showed that these early "vertical" wells were in fact far from being vertical.

A well that was not vertical was at first considered a disadvantage, for the following reasons.

(a) It meant that more footage than necessary had to be drilled in order to reach the producing zone. A slanted or deviated drill therefore was more time-consuming and expensive to drill than a straight vertical well.

(b) The true vertical depth of the producing zone could not be determined accurately in a deviated wellbore, and so made the planning of future wells more difficult.

(c) The crooked shape of the wellbore increased the wear on the drill string, making a failure more likely. The deviated wellbore also made any subsequent fishing job more difficult.

(d) If the wellbore crossed a lease boundary, the operator could be prosecuted for trespassing into another operator's acreage. There were in fact several court cases arising from this problem. This led to "deviation clauses" being written into drilling contracts, which imposed some limit (5° or less) on the deviation that could be permitted. There was therefore a legal as well as practical necessity to control the path of the wellbore.

The earliest application of directional drilling was probably to sidetrack

1

around a fish. If the obstruction at the bottom of the hole could not be retrieved, the driller would have to drill around it. As early as 1895 special tools and techniques were being used to do this. The first recorded instance of a well being deliberately deviated to reach its objective was at Huntington Beach, California, in the early 1930s. At that time it was common practice to produce oil from beneath shallow coastal waters by setting up a drilling rig on a jetty that ran out at right angles to the shore. An enterprising driller decided on an alternative plan, which was to set up the rig on the land and drill a slanted hole out under the sea-bed. This was the beginning of directional drilling as it is known today.

Further applications of directional drilling were soon realized. In 1934 a deviated well was drilled to kill a blow-out on the Conroe Field in East Texas. The rig was set up some distance away and the relief well drilled to intersect the wild well. The blow-out was killed by pumping heavy mud down the relief well.

The drilling of relief wells demanded better directional control and monitoring. Directional surveying instruments known as "single shots" allowed the operator to plot the course of the well much more accurately. Deflecting tools such as whipstocks could also be oriented in the required direction. The early single shots, however, relied on a magnetic compass that could be affected by local magnetic fields in the drill string or casing. Non-magnetic drill collars and gyroscopic surveying tools were introduced to improve accuracy under these circumstances.

By the end of World War II the increased demand for petroleum led to more exploration into remote and hostile areas. Large deposits of oil and

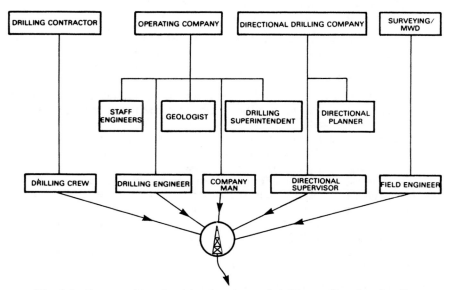

Fig. 1.1. Personnel involved in planning and drilling a directional well.

gas were located under the sea, but the high cost of drilling offshore wells prohibited their exploitation. The number of individual drilling platforms could be reduced by drilling directional wells from one central platform. Many offshore fields would not be economically viable without directional wells. Offshore developments led to a big expansion in the use of directional drilling. The continuing need to reduce drilling costs provided the incentive to produce new tools and techniques to improve efficiency. During the past 20 years many innovations have taken place, including computerized well planning, more use of downhole motors and turbines, techniques to drill horizontal wells and the introduction of measurement while drilling (MWD).

Directional drilling has now become an essential element in oilfield

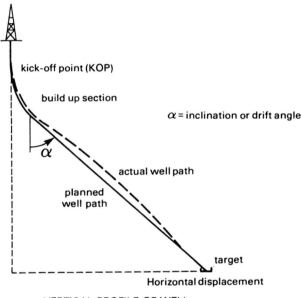

VERTICAL PROFILE OF WELL

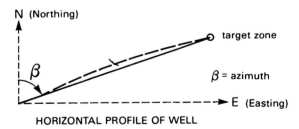

HORIZONTAL PROFILE OF WELL

Fig. 1.2. Vertical and horizontal views of a directional well.

development, both onshore and offshore. Despite the advances made in drilling technology, there is still a great need for personnel with the proper training and experience to use the technology to its maximum benefit. Directional drilling service companies have built up the required level of expertise. An operating company will most likely call upon these companies to assist in the planning of the well and to provide a directional driller to supervise operations at the well-site. The directional driller is responsible for directing the well along the proposed wellpath and successfully intersecting the target. The full range of personnel involved in drilling a directional well is shown in Fig. 1.1. As well as those working on the rig, there are many involved at the planning stage and in monitoring the progress of the well. The importance of planning should not be underestimated.

Directional drilling has made great advances over a relatively short period of time. It is now fairly common to drill a well as shown in Fig. 1.2, with a horizontal displacement of 2–3 miles and a true vertical depth of over 10,000 ft. The improvement in tools and techniques is continuing and will allow more efficient exploitation of petroleum reserves.

FURTHER READING

"Technical advances broaden use of highly deviated and horizontal drilling methods", *Journal of Petroleum Technology*, February 1981.
History of Oil Well Drilling, J. E. Brantly, Gulf Publishing Company, 1971.

Chapter 2

APPLICATIONS OF DIRECTIONAL DRILLING

The applications of directional drilling can be grouped into the following categories:

(a) sidetracking;
(b) drilling to avoid geological problems;
(c) controlling vertical holes;
(d) drilling beneath inaccessible locations;
(e) offshore development drilling;
(f) horizontal drilling;
(g) non-petroleum uses.

SIDETRACKING

During the drilling of a well, an obstruction (or fish) may become stuck at the bottom of the hole. This may be the result of a drill string failure or an intentional back-off where the lower part of the string is left in the hole. No further progress can be made if the fish cannot be pulled out of the hole. In the early days of rotary drilling it was soon realized that it was much cheaper to drill around the obstruction rather than abandon the hole and start again.

A cement plug is placed on top of the fish and is allowed to set firmly. This forms a good foundation from which the new section of hole can be kicked off. A whipstock was the first tool designed to deflect the wellbore around a fish, but a downhole motor and bent sub are more likely to be used today. The bent sub can be oriented in the required direction by using MWD or a steering tool that will provide continuous monitoring of the wellpath. Once the sidetrack has been drilled around the obstruction, the hole is continued down to the target (see Fig. 2.1).

Sidetracking may also be carried out for a re-drill or re-completion. If

5

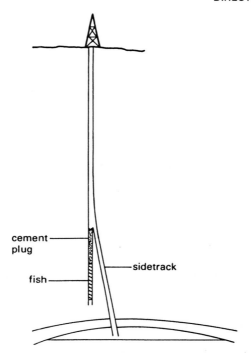

Fig. 2.1. Sidetracking around a fish.

the original well did not locate the anticipated formation or is producing from a zone that has become depleted, the hole can be plugged back and then sidetracked towards a new target. If the kick-off point is in a section of cased hole, a window must be milled out of the casing to allow the sidetrack to be drilled. The same principle can be employed in an exploration well to test several different zones using the same wellbore.

DRILLING TO AVOID GEOLOGICAL PROBLEMS

Petroleum reservoirs are sometimes associated with salt dome structures. Part of the salt dome may be directly above the reservoir, so that a vertical well would have to penetrate the salt formation before reaching the target. Drilling through a salt section introduces certain drilling problems such as large washouts, lost circulation, and corrosion. In this situation it would be wiser to avoid the salt formation by drilling a directional well as shown in Fig. 2.2.

If a well is drilled vertically through a steeply dipping fault plane there is a risk of movement or slippage along that plane. This problem can also be avoided by drilling a directional well.

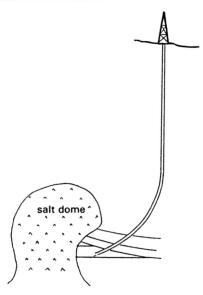

Fig. 2.2. A directional well drilled beneath a salt dome.

CONTROLLING STRAIGHT WELLS

To keep vertical wells on target and prevent them from straying across lease boundaries, directional techniques have to be used. Small deviations from the planned course can be corrected by altering certain drilling parameters or changing the bottom hole assembly (BHA). More serious deviations may require the use of a downhole motor and bent sub to make a correction run or drill a sidetrack. The same problem may occur in the tangential section of a directional well.

INACCESSIBLE LOCATIONS

Oilfields are often located directly beneath natural or man-made obstructions. Permission may not be granted to drill in some sensitive areas, since there may be a risk to the environment. In such cases, it may be possible to exploit the reserves by drilling directional wells from a surface location outside the restricted area (Fig. 2.3).

When a blow-out destroys or damages the rig in such a way that capping operations are impossible, relief wells are drilled to bring the blow-out safely under control. Improved directional techniques have enabled relief wells to reach targets less than 10 ft from the blow-out. Often two relief wells are drilled simultaneously from different surface locations to ensure that the blow-out is killed.

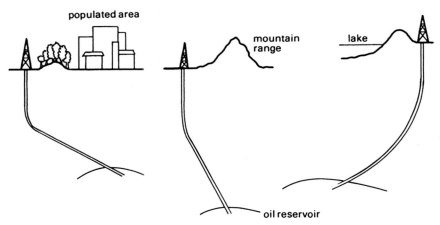

Fig. 2.3. Examples of directional wells drilled beneath inaccessible locations.

OFFSHORE DEVELOPMENT DRILLING

One of the major applications of directional drilling over the past 20 years has been the development of offshore reservoirs. Many oil and gas deposits in the Gulf of Mexico, North Sea and other areas are situated beyond the reach of land-based rigs. To drill a large number of vertical wells from individual platforms is clearly very expensive and impractical. The conventional approach for a large oilfield has been to install a fixed platform on the seabed, from which several directional wells may be drilled. The bottom hole locations can be carefully spaced for optimum recovery. All the necessary production facilities can also be centralized on the platform, from which the oil can be exported via pipelines or tankers. In the hostile conditions of the North Sea, drilling and production platforms have been installed in over 500 ft of water and over 100 miles from the nearest land. More than 50 wells may be directionally drilled from some of these large platforms (Fig. 2.4).

In a conventional development, the wells cannot be drilled until the platform has been constructed and installed in position. This may mean a delay of 2–3 years before production can begin. This will have a very detrimental effect on the overall economics of the development, especially for the smaller marginal fields. To reduce this delay in starting production, some of the development wells can be pre-drilled through a subsea template while the platform is being constructed. These wells will be directionally drilled from a semi-submersible rig and tied back to the platform once it has been installed. (Fig. 2.5).

The template is lowered down to the seabed and secured in position by piles. The development wells are then drilled through the template using equipment similar to that used in exploratory drilling from a floating rig (subsea wellheads and BOP stacks). The major difference between drilling

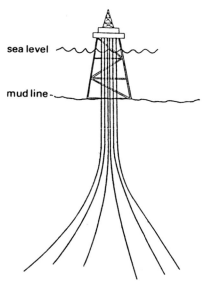

Fig. 2.4. Development wells drilled from a fixed platform.

deviated wells from a floater, as opposed to a fixed platform, is the effect of the vessel's movements. A heave compensator is essential to ensure that constant WOB can be applied and the pipe can be controlled while tripping out through sticky formations. Drilling operations may have to be suspended completely in very bad weather.

As exploration moves into deeper waters (1000 ft and more), the cost of

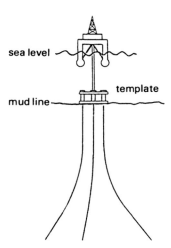

Fig. 2.5. Development wells drilled from a floating rig through a subsea template.

a fixed platform rises sharply. Many operators are looking towards floating drilling and production units for these deep-water developments (e.g. tension leg platforms). Again, these fields will rely on directional drilling from the floating platform for efficient exploitation of the reserves.

HORIZONTAL DRILLING

Conventional directional wells may be drilled to an inclination of around 60°. Inclinations beyond 60° give rise to many drilling problems that substantially increase the cost of the well. However, there are certain advantages in drilling highly deviated wells and horizontal wells. These include:

(a) increasing the drainage area of the platform;
(b) prevention of gas coning or water coning problems;
(c) increased penetration of the producing formation;
(d) increasing the efficiency of enhanced oil recovery (EOR) techniques;
(e) improving productivity in fractured reservoirs by intersecting a number of vertical fractures.

The extra cost of drilling a horizontal well must be justified by the increased productivity it will provide. The potential benefits and the risk involved must be carefully considered before drilling the well. Normal drilling procedures may have to be modified and special drilling equipment may have to be installed to drill and complete a horizontal well (see Fig. 2.6). Horizontal drainholes (short radius drilling) can also be applied to overcome certain reservoir problems (Fig. 2.7).

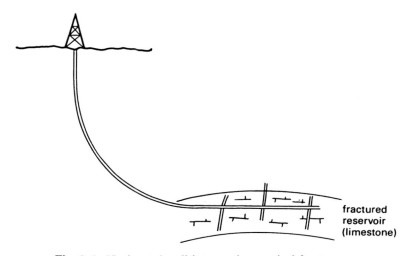

fractured
reservoir
(limestone)

Fig. 2.6. Horizontal well intersecting vertical fractures.

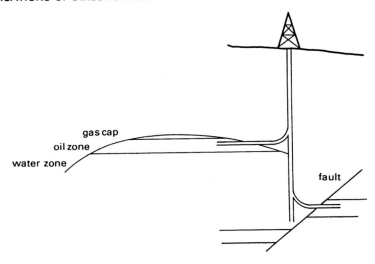

gas cap
oil zone
water zone

fault

Fig. 2.7. Horizontal drainhole.

NON-PETROLEUM APPLICATIONS

Mining industry
The drilling of small-diameter boreholes in rock to measure thickness of the strata and to obtain core samples is well established. Indeed, some of the techniques used in the oil industry were adopted from earlier techniques used in mining (e.g. borehole surveying to measure inclination and direction). Directional wells are also used to produce methane gas that is contained in coal seams. The methane presents a safety hazard and must be drained off before mining operations can begin. In deep coal seams that are beyond the reach of conventional mining techniques, directional wells have been drilled for *in situ* gasification projects.

Construction industry
An unusual application of directional drilling is the installation of pipelines beneath river beds. A small-diameter pilot hole is drilled in a smooth arc beneath the river until it emerges on the other side. This acts as a guide for the larger-diameter pipe that forms the conduit. The pilot hole is drilled using a downhole motor and bent sub. The hole is drilled through soft sediments about 40 ft below the river bed. The technique has been used to cross rivers up to 200 ft wide.

Geothermal energy
In certain areas of the world the high geothermal gradient found in some rocks can be harnessed to provide energy. The source rock (e.g. granite) is generally impermeable except for vertical fractures. Extracting the heat from this rock requires the drilling of injection and production wells. The wells are directionally drilled to take advantage of the orientation of the

fractures. The high temperatures and hardness of the rock cause some major drilling problems (such as severe abrasion of downhole components, reduction in yield strength of steel at temperatures greater than 200°C, and the need for special downhole motors).

FURTHER READING

"Offshore deviated wells—pre-drilling on template", M. Chatlain, *Petrole Informations*, 2 April 1981.
"Directional drilling adapted for pipe line crossings", *Pipe Line Industry*, September 1974.
"Onshore and offshore European horizontal wells", L. H. Reiss, A. P. Jourdan, F. M. Giger and P. A. Armesson, O.T.C. paper 4791.
"Directional drilling equipment and techniques for deep, hot granite wells', T. L. Brittenham, J. W. Neudecker, J. C. Rowley and R. E. Williams. S.P.E. paper no. 9227.
"Improved directional drilling will expand use", W. J. McDonald, W. A. Rehm and W. C. Maurer, *Oil and Gas Journal*, 26 February 1979.

Chapter 3

DEFLECTION TOOLS AND TECHNIQUES

When a vertical well is being drilled, the axis of the borehole will not be truly vertical. There will always be a tendency for the bit to deviate from its intended course owing to a combination of formation effects and the behaviour of the bottom hole assembly. This is a well-known problem in drilling straight holes and can now be controlled by better knowledge of directional behaviour and the use of deflection tools.

One of the major problems in straight and deviated holes is how to predict the amount of deviation as the well is being drilled. Experience of other wells drilled in the same area through similar formations is obviously useful. A careful record should be kept of how each BHA behaved through each interval. In drilling from multiwell platforms, this record can be referred to when selecting a BHA in the next well. Since 1970 many efforts have been made to predict the behaviour of a BHA by analysing the various forces which are acting on the bit. Several companies now have drilling models based on computer analysis techniques (including finite-element analysis). To be effective these computer models must take account of the many factors involved in the drilling process.

NATURAL FORMATION EFFECTS

The formations encountered when drilling oil wells are very rarely homogeneous and isotropic. One is more likely to find a sequence of different layers, with each layer having its own drillability characteristics. The bit may have to drill through alternating layers of hard and soft rock. Furthermore, these strata will probably not be lying evenly in horizontal beds but will be dipping at some angle. The geology may be further complicated by faulting and folding of the strata.

As the bit drills across a formation boundary it will tend to be deflected from its original course. Experience has shown that where the formations

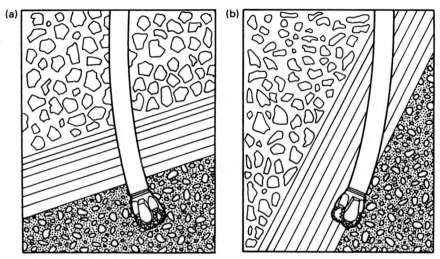

Fig. 3.1. Effect of formation dip on well path. (a) Bit drilling up dip where formations are dipping at angles less than 45°. (b) Bit drilling up dip where formations are dipping at angles greater than 60°.

are steeply dipping (greater than about 60°) the bit tends to drill parallel with the bedding planes. Where the formation dip is less steep, the bit will tend to drill at right angles to the bedding planes (Fig. 3.1). In addition to changes in inclination there may also be changes in direction where the bit tends to "walk". Under normal rotary drilling, the bit will tend to walk to the right, but with a downhole motor the effect of reactive torque may force it to the left. Various theories have been proposed to explain these formation effects, but it is difficult to predict the extent of the deviation without detailed geological knowledge, for example of dip angles, presence of faults and the relative hardness of the formations. Drilling parameters such as WOB and RPM, and hydraulics, will also affect the amount of deviation.

MECHANICAL FACTORS

The physical properties of the various downhole components of the BHA have a significant effect on how the bit will drill. It is the bottom 100–300 ft of the BHA that has the greatest influence on behaviour. This is sometimes referred to as the "active length" of the drill string and is made up of drill collars, subs and stabilizers.

Drill Collars

The primary function of the drill collars is to provide sufficient WOB (Weight on bit). The weight of the collars also ensures that the drill pipe is kept in tension to prevent buckling. Since the collars are under compression, they will tend to bend under the applied loading. The amount of bending will depend on the material and the dimensions of the collar. The stiffness of the collar is the product of the collar's moment of inertia (I) and the modulus of elasticity for that material (E). The polar moment of inertia for a hollow cylinder is given by

$$I = \frac{\pi}{64}(D^4 - d^4)$$

where I = moment of inertia (in.^{4})
  D = outside diameter (in.)
  d = inside diameter (in.).

The modulus of elasticity for various materials can be obtained from manufacturer's specifications; e.g. for steel $E = 29 \times 10^6$ psi; for aluminium $E = 11 \times 10^6$ psi; for monel $E = 26 \times 10^6$ psi. An aluminium drill collar therefore will be more flexible than a steel drill collar of similar dimensions. The WOB supplied by the drill collars will depend on the density of the material and the collar's dimensions. Although thick-walled collars will add more weight, this will also mean a stiffer assembly that may not be suitable for a directional well in which differential sticking is likely to occur. Spiral drill collars with external grooves cut into their profile may be used to reduce the contact area between the BHA and the formation (Fig. 3.2(a)). It should be remembered that in deviated holes the total weight of the drill collars will not all be applied to the bit. Much of the weight will be applied to the borehole walls. The actual WOB is a function of cos α, where α is the angle of inclination.

Heavy-weight Drill Pipe (HWDP)

A joint of HWDP has a greater wall thickness and longer tool joints than normal drill pipe. Midway between the tool joints is an integral wear pad which acts as a stabilizer (Fig. 3.2(b)). HWDP has basically the same functions as a drill collar but has much less contact area with the formation. Like the drill collars, HWDP can be run in compression. The use of HWDP in a directional well will therefore

 (i) reduce torque and drag on the drill string;
 (ii) reduce the risk of differential sticking;
(iii) reduce the risk of tool joint failures when drilling through dog-legs.

The BHA in a directional well may have 20 or more joints of HWDP between the drill collars and drill pipe.
   The effect of bending can be assessed by calculating the section modulus for adjacent drill string components (see Table 3.1).

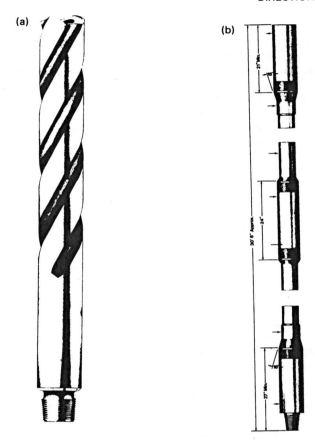

Fig. 3.2. (a) Spiral drill collar. (b) Heavy weight drill pipe (courtesy of Smith International Inc.).

$$\text{Section modulus} = \frac{\text{moment of inertia } (I)}{\text{external radius of tube}}$$

The ratio of section moduli should be less than 5.5 to reduce damage due to bending.

Stabilizers

Stabilizers are fairly short subs which have blades attached to their external surface. By providing support for the BHA at certain points they can be used to control the trajectory of the well. The blades can be either straight or spiral in shape. Spiral blades can give 360° contact with the borehole. Various different types of stabilizers are available (Fig. 3.3).

TABLE 3.1 Properties of Drill String Components

Note that the ratio of section moduli between adjacent components should be less than 5.5 to avoid bending stresses in the string.

	Outside diameter (in)	Inside diameter (in)	Weight/ft (lb/ft)	Section modulus (in³)
	3.5	2.992	9.5	2.0
	3.5	2.602	15.5	2.9
	4.5	3.826	16.6	4.3
Drill pipe	4.5	3.640	20.0	5.1
	5.0	4.276	19.5	5.7
	5.0	4.0	25.6	7.2
	5.5	4.778	21.9	7.0
	5.5	4.670	24.7	7.8
	3.5	2.062	25.3	3.7
Heavy weight drill pipe	4.0	2.562	29.7	5.2
(Range II)	4.5	2.750	41.0	7.7
	5.0	3.0	49.3	10.7
	6.0	2.812	75.0	20.2
	6.0	2.25	82.6	20.8
	6.75	2.812	100.4	29.3
Drill collars	6.75	2.25	108.5	29.8
	8.0	3.0	147.0	49.3
	8.0	2.812	149.6	49.5
	9.5	3.25	212.8	83.0
	9.5	3.0	217.0	83.3

(a) Welded blade. Steel blades are welded on to the body of the stabilizer. The mud in the annulus flows between the blades. The blades make contact with the wall and may cause hole enlargement in soft formations. This type of stabilizer can be used when the gauge size remains constant.

(b) Integral blade. These are more expensive than welded blade type stabilizers, since they are machined from one piece of metal. They are generally used to provide a larger contact area. Tungsten carbide can be used on the blades to provide better wear resistance in more abrasive formations.

(c) Sleeve stabilizers. These consist of replaceable sleeves that are mounted on the stabilizer body. They offer the advantage of changing out a sleeve with worn blades or replacing it with one of another gauge size. The blades can be dressed with tungsten carbide inserts for abrasive formations.

(d) Non-rotating stabilizers. These stabilizers are used to centralize the drill collars, but the rubber sleeve allows the string to rotate while the sleeve remains stationary. The wear on the blades is therefore much less than in other stabilizers and so they can be used in harder formations.

Stabilizers can be installed just above the bit (near-bit stabilizers) or at any point within the BHA (string stabilizers). Two stabilizers can also be

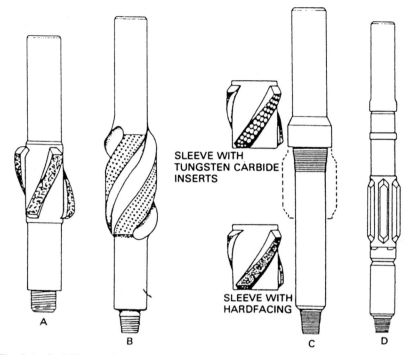

SLEEVE WITH
TUNGSTEN CARBIDE
INSERTS

SLEEVE WITH
HARDFACING

A

B   C   D

Fig. 3.3. Stabilizers: (a) welded blade; (b) integral blade; (c) sleeve; (d) non-rotating sleeve (courtesy of Eastman-Christensen).

run in tandem if necessary ("piggy-back"). Stabilizers are inserted at drill collar connections. This limits their spacing to 30 ft or multiples of 30 ft. Closer spacing can be achieved by using shorter drill collars (pony collars) that are 10–15 ft long. "Clamp on" stabilizers can be used to provide support at some point along the length of a collar. Any stabilizer that is placed near a magnetic surveying tool must be made of non-magnetic material, to prevent distortion of the survey results.

Roller Reamers

Roller reamers may be used as an alternative to stabilizers in certain circumstances. Their purpose is to centralize the BHA and ensure that full gauge hole is drilled in abrasive formations. The roller reamer has a number of cutters which make contact with the formation (Fig. 3.4). A near-bit roller reamer is often used to prevent bit walk.

FORCES ACTING ON THE BIT

In a deviated well, the drill collars will make contact with the low side of the hole. If no stabilizers are included in the BHA (i.e. a 'slick' assembly)

Fig. 3.4. Roller reamer.

the collars will make contact with the borehole at a distance L from the bit (Fig. 3.5). The distance L is known as the tangent length. The unsupported length of collars below the tangent point creates a pendulum effect that exerts a side force at the bit. The maximum side force can be determined from:

$$F = \frac{LW_c \sin \alpha}{2}$$

where F = maximum side force (lb)
   L = tangent length (ft)
   W_c = weight per unit length of collars (lb/ft)
   α = angle of inclination.

Since this force is tending to decrease the inclination of the hole, it is referred to as a negative side force. As WOB is applied, the tangent point is lowered, thus reducing the side force. The bending of the drill collars at the bit may also result in a component of the axial load being applied to increase the hole angle (positive side force). With increasing WOB, therefore, the negative side force reduces as the positive side force becomes greater. The vector sum of the resultant side force and the axial force should determine the deviation of the hole, but in reality the anisotropy of the formation must also be considered.

The placement of a stabilizer in the BHA will affect the size of the side

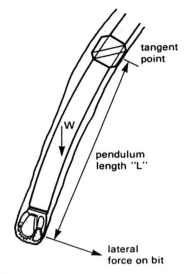

Fig. 3.5. Pendulum effect causing a drop in inclination due to weight of drill collars.

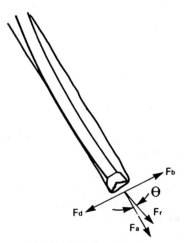

Fig. 3.6. Resolving axial load and resultant side force.
F_b = building force (positive);
F_d = dropping force (negative);
F_a = axial force;
F_r = resultant force.

force, and hence will dictate whether the BHA will build or drop angle. A stabilizer placed immediately above the bit will act as a pivot or fulcrum. The weight of the collars above the stabilizer will act as a lever to make the bit build angle. As the distance between the bit and this stabilizer increases, the upward force on the bit is reduced. If placed higher up the string, the stabilizer will determine the tangent length and produce a pendulum effect at the bit.

When the hole angle is to be maintained, a holding assembly must be run. This assembly will cancel out the fulcrum and pendulum effects. More stabilizers are used to provide extra stiffness to resist bending.

ROTARY ASSEMBLIES

A rotary assembly is a BHA which is driven solely by the rotary table at surface. No downhole motors or turbines are included. By careful placement of stabilizers, rotary assemblies can be designed to build, hold or drop the angle of inclination.

Building Assembly

This type of assembly is usually run in a directional well after the initial kick-off has been achieved by using a deflection tool. A single stabilizer placed above the bit will cause building owing to the fulcrum effect. The addition of further stabilizers will modify the rate of build to match the required well trajectory. If the near-bit stabilizer becomes undergauge, the side force reduces. Typical building assemblies are shown in Fig. 3.7. Assemblies A and B respond well in soft or medium formations. The inclusion of an undergauge stabilizer in assembly C will build slightly less angle. By bringing the second stabilizer closer to the near-bit stabilizer, the building tendency is increased. In hard abrasive rocks, the problems of bit wear are significant. To maintain gauge hole, the near-bit and second stabilizer should be replaced by roller reamers. The build rate should be kept below 2° per 100 ft to reduce the risk of dog-legs. The amount of WOB applied to these assemblies will also affect their building characteristics. Too much WOB will cause rapid build-up of angle.

Holding Assemblies

Once the inclination has been built to the required angle, the tangential section of the well is drilled using a holding assembly. The object here is to reduce the tendency of the BHA to build or drop angle. In practice this is difficult to achieve, since formation effects and gravity may alter the hole angle. To eliminate building and dropping tendencies, stabilizers should be placed at close intervals, using pony collars if necessary. Assembly D in Fig. 3.7 has been used successfully in soft formations. The undergauge stabilizer in assembly E builds slightly to counter gravity. In harder formations the near-bit stabilizer is replaced by a reamer. Generally only three stabilizers should be used, unless differential sticking is expected. Changes in WOB

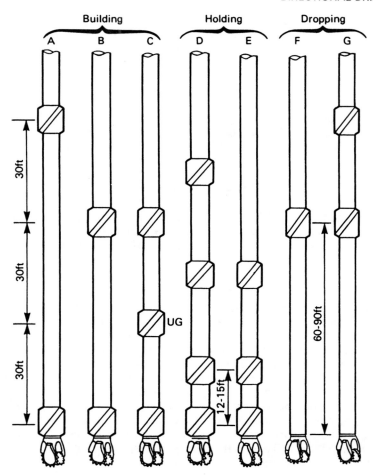

Fig. 3.7. Bottom hole assemblies.

will not affect the directional behaviour of this type of assembly, and so optimum WOB can be applied to achieve maximum penetration rates. A packed hole assembly with several stabilizers should not be run immediately after a downhole motor run.

Dropping Assemblies

In directional wells, only an S shape profile requires a planned drop in angle. The other application of a dropping assembly is when the inclination has been increased beyond the intended trajectory and must be reduced to bring the well back on course. It is best to drop angle in a section of softer formation, since the response to a pendulum type assembly in hard rock is very slow. Figure 3.7 gives some typical dropping assemblies (F and G).

These are more effective in high-angled holes. If hole angle does not reduce, the WOB can be reduced, although this will also reduce the penetration rate.

DEFLECTING TOOLS

Although rotary assemblies can be designed to alter the path of the wellbore, there are certain circumstances in which it is necessary to use special tools (e.g. kicking-off and sidetracking).

Whipstocks

The use of a whipstock to sidetrack around a stuck fish dates back to the late 1890s. The original whipstock was probably a wedge-shaped piece of wood that was dropped down on top of the fish. When the bit was lowered down the hole the tapered side of the wedge deflected the bit away from the fish. A new section of hole could then be drilled at a slight angle to the vertical. Whipstocks were later used to kick-off directional wells with the aid of directional surveying instruments to check the orientation of the tapered edge. The direction in which the tapered edge was facing became known as the "toolface". There are several different types of whipstock available.

A "removable whipstock" can be used to initiate deflection in open hole, or straighten vertical wells that have become crooked. The whipstock consists of a steel wedge with a chisel-shaped point at the bottom to prevent movement once drilling begins. The tapered concave section has hard facing to reduce wear. At the top of the whipstock is a collar that is used to withdraw the tool after the first section of hole has been drilled. The whipstock is attached to the drill string by means of a shear pin. Having run into the hole, the drill string is rotated until the toolface of the whipstock is correctly positioned. By applying weight from surface, the chisel point is set firmly into the formation or cement plug. The retaining pin is sheared off and drilling can begin (see Fig. 3.8a).

A small-diameter pilot hole is drilled to a depth of about 15 ft below the toe of the whipstock. After this rathole has been surveyed, the bit and whipstock are tripped out. A hole opener is then run to ream out the rathole to full size. Once the deflected section of hole has been started, a rotary building assembly can be run to continue the sidetrack.

If there is a build-up of cuttings at the bottom of the hole, it may be difficult to position the whipstock properly. This led to the introduction of the "circulating whipstock", which contains a passageway to allow mud to wash out these cuttings or fill from the bottom of the hole (see Fig. 3.8b).

A "permanent whipstock" is used mainly in cased hole for sidetracking around a fish or by-passing collapsed casing. A casing packer is set at the kick-off point to provide a base for the whipstock. The whipstock is run with a mill that will cut a "window" in the casing. After setting the whipstock in the required direction and shearing the retaining pin, the

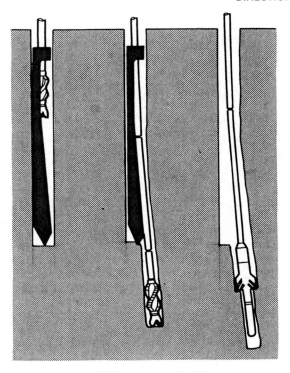

Fig. 3.8. (a) Operation of whipstock.

milling operation begins. Once the window has been cut, the mill is replaced by a small diameter pilot bit. The pilot hole is subsequently reamed out to full size.

If used correctly the whipstock is a reliable and effective deflecting tool. It is able to provide a controlled and gradual build up of angle. There are however several disadvantages.

(a) The rathole that is drilled initially must be reamed out, requiring a new BHA to be run.
(b) While drilling, the whipstock may rotate, deflecting the bit away from its intended direction.
(c) As the bit drills off the whipstock, some drop in inclination may occur.
(d) When using a permanent whipstock to mill a window in the casing, the window itself is often too small. It may be advisable to use a section mill to cut out a larger length of casing, then set a cement plug and deflect the wellbore with a mud motor and bent sub.

As a result of these disadvantages the whipstock has now largely been replaced by other deflection techniques. It may still be used for specific

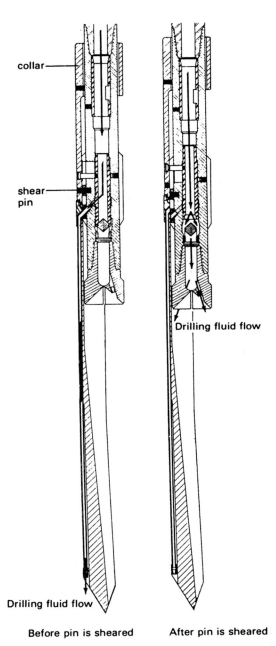

collar

shear pin

Drilling fluid flow

Drilling fluid flow

Before pin is sheared   After pin is sheared

Fig. 3.8. (b) Circulating whipstock (courtesy of Eastman-Christensen).

applications in which other methods have been unsuccessful or cannot be attempted (e.g. when high temperatures prevent the use of a mud motor).

Jet Deflection

Jet deflection is a technique best suited to soft–medium formations in which the compressive strength is relatively low. The hydraulic power of the drilling fluid is used to wash away a pocket of the formation and initiate deflection. A specially modified bit must be used that has one nozzle much larger than the other two. A two-cone bit with a large "eye" may also be used (Fig. 3.9a). The bit is run on an assembly which includes an orienting sub and a full-gauge stabilizer near the bit (see Fig. 3.9b). Once on bottom, the large nozzle is oriented in the required direction. Maximum circulation rate is used to begin washing without rotating the drill string. The pipe is worked up and down while jetting continues, until a pocket is washed away. At this stage the drill string can be rotated to ream out the pocket and continue building angle as more WOB is applied. Surveys must be taken frequently to ensure the inclination and direction are correct. If it is found that the deflected section of the well is not following the planned trajectory, the large nozzle can be re-oriented and jetting can be repeated. This method has several advantages.

(a) A full gauge hole can be drilled from the beginning (although a pilot hole may be necessary in some cases).
(b) Several attempts can be made to initiate deflection without pulling out of the hole.

There are associated disadvantages.

(a) The technique is limited to soft–medium formations (in very soft rocks too much erosion will cause problems).
(b) Severe dog-legs can occur if the jetting is not carefully controlled (if the drilling is fast, surveys must be taken at close intervals).
(c) On smaller rigs there may not be enough pump capacity to wash away the formation.

Under suitable geological conditions and with good directional monitoring, jet deflection is a very cost-effective means of kicking-off directional wells.

Rebel Tool

The "rebel tool" is primarily used to change the azimuth of the trajectory. It is normally run in the tangential section of the hole to correct the tendency of the bit to walk to the left or right. The rebel tool is run with a conventional rotary assembly and should be placed immediately above the bit for maximum effect.

The body of the tool is similar to a short drill collar 8–16 ft long. The tool is available in diameters ranging from $4\frac{13}{16}''$ to $8\frac{7}{8}''$. The lower end of the tool has a box connection looking down so that it can be made up directly on top of the bit. Along the length of the collar is a recess that accommo-

(a)

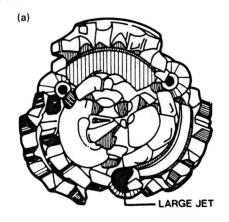

LARGE JET

(b)

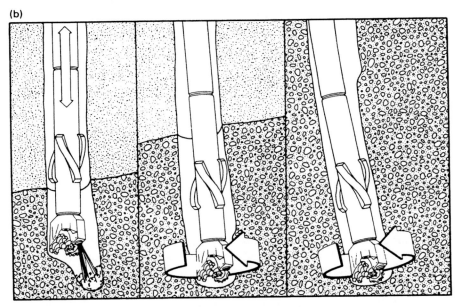

Fig. 3.9. (a) Jet deflection bit (courtesy of Eastman-Christensen). (b) Jet deflection technique.

dates a shaft or torsion rod (Fig. 3.10). At each end of the shaft is a curved paddle, and it is this mechanism that deflects the bit. While drilling a directional well, the top paddle rotates to the low side of the hole and is forced into the recess. This causes the shaft to turn and forces the lower paddle against the borehole, exerting a lateral force on the bit. The paddles can be fitted to provide a left- or right-hand turn at the bit. The paddles

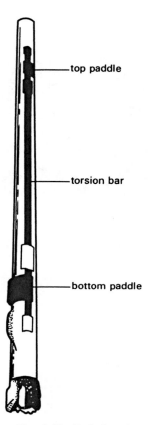

Fig. 3.10. Rebel tool.

have tungsten carbide hard-facing to ensure high wear resistance. Surveys should be taken at fairly close intervals to monitor direction. The advantages of using the rebel tool are that:

(a) It saves time, since no orientation is required, and only minor changes to drilling parameters are necessary (rpm reduced to give more torque at the paddles; circulation rate increased to clean paddles).
(b) It is much cheaper to run than a downhole motor and bent sub.

The rebel tool will provide a gradual change in direction ($\frac{1}{2}$–$2\frac{1}{2}$° per 100 ft), but it requires gauge hole to work properly.

Downhole Motor and Bent Sub

The most common deflection technique in current use involves running a positive displacement motor to drive the bit without rotating the drill string. The deflection is provided by a special sub placed above the motor to create a side force at the bit.

The bent sub is a short length of drill collar usually about 2 ft long. The axis of the lower (pin) connection is machined slightly off-vertical. The amount of offset angle can vary between ½° (for very gradual changes of trajectory) and 3° (for very rapid changes). The bent sub forces the bit and motor to drill in a specific direction that is determined by the tool face. The toolface is indicated by a scribe line marked on the inside of the bend in the sub (Fig. 3.11). The amount of deflection is a function of the offset

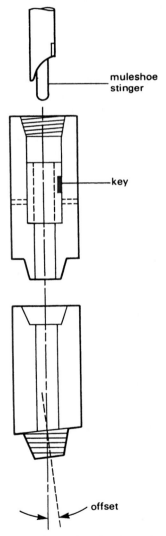

muleshoe stinger

key

offset

Fig. 3.11. (above) Orienting sub.
Fig. 3.12. (below) Bent sub.

provided by the bent sub, the stiffness of the downhole motor and the hardness of the formation.

Once the assembly is run to bottom, the orientation of the bent sub can be measured by surveying tools. The muleshoe key of the orienting sub is aligned with the scribe line, so that when the survey tool is seated it will give the direction of the tool-face. The bent sub itself may contain an orienting sleeve, or a separate orienting sub may be run immediately above the bent sub (Fig. 3.12). A typical deflecting assembly is shown in Fig. 3.13. Without rotating the drill string, mud is pumped through the drill string to operate the motor and drive the bit. The reactive torque of the motor may cause the drill string to twist to the left as drilling proceeds. To compensate for this effect the bent sub should initially be oriented towards the right of the required direction. An MWD tool or steering tool should be run to monitor the toolface heading continuously.

For difficult deflections (e.g. deflecting through a casing window) a bent housing can be installed within the motor itself. This is a special device that is placed between the stator and bearing assembly to give a slight bend of 1–1.5°. The bent housing introduces the deflection much closer to the bit than is possible with a bent sub on top of the motor. This means that a bent housing will provide a larger turn than a bent sub of similar size.

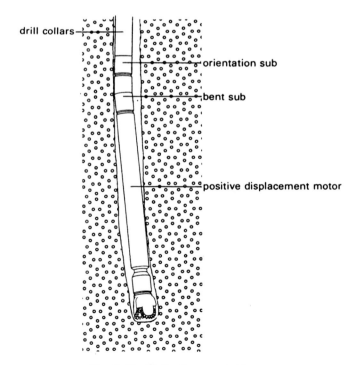

Fig. 3.13. Deflection assembly.

The advantages of using a bent sub or bent housing with a mud motor are that:

(a) Full-gauge hole can be drilled without the need for a pilot hole.
(b) The continuous side force produced at the bit by the bent sub gives a smooth curvature with less risk of severe dog-legs.
(c) Depending on the orienting of the bent sub, this technique can be used to build or drop inclination, and to steer the bit to the left or right.

When a very rapid change of angle is required, a bent sub can be used together with a bent housing. One important disadvantage is that the rubber components of the motor can be damaged by high temperatures and cannot be used in holes in which the temperature is 280°F or greater. Bit life may also be reduced, owing to the faster rotational speed of the downhole motor.

Downhole Turbines

A downhole turbine can be used in the same way as a positive displacement motor (PDM) to deflect the wellbore. The shorter length of the PDM, however, gives a greater advantage over the turbine for kicking-off and sidetracking operations. In the long tangential section of the directional well, however, turbines may be more cost-effective than conventional rotary methods. It is possible to steer the turbine over this section by means

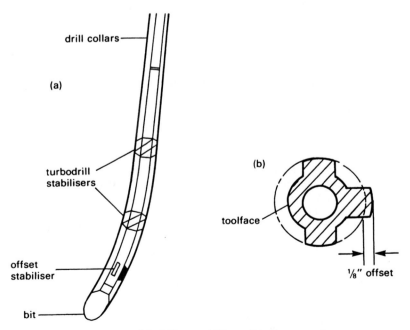

Fig. 3.14. (a) Offset stabilizer. (b) Cross-section.

of an offset stabilizer (Fig. 3.14). The offset stabilizer can be oriented in the required direction, the drill string is not rotated, and the turbine drives the bit along the desired course. Once the well is brought back onto the planned trajectory, the drill string can be rotated. By rotating the offset stabilizer a slightly overgauge hole will be drilled.

ORIENTATION OF DEFLECTING TOOLS

When planning a trajectory change, it is essential to have the deflecting tool pointing in the required direction before drilling begins. Although the bent sub or whipstock can be turned on surface, actual downhole orientation must be checked by surveying instruments.

Toolface Setting

To set the toolface properly the directional driller must have certain information:

(a) the present inclination and azimuth of the hole;
(b) the required change in inclination or azimuth to correct the trajectory or kick off the well;
(c) the expected rate of change (i.e. the dog-leg) that the deflecting tool can provide.

A graphical method can be used to calculate the toolface setting. The following example can be used to illustrate how the method works.

Example
During a kick-off operation a new toolface setting is required to bring the well back on course. The present depth is 4000 ft, inclination = 8° and azimuth = 050° (N 50° E). The well must be turned 10° to the right. With a downhole motor and bent sub the expected overall change in angle is $1\frac{1}{2}$° per 100 ft. How should the toolface be oriented?

The solution can be represented on a Ragland diagram as shown in Fig. 3.15. The line OA represents the present direction (050°). Along this line the current inclination (8°) is scaled off. This fixes the position P. At this point a circle representing the expected change of angle ($1\frac{1}{2}$°) is drawn. The line OB represents the new direction (060°) which cuts the circle at 2 points X and Y. The angle $\widehat{APY}$ gives the orientation angle. Toolface settings are usually given in terms of degrees left or right from the High Side of the hole. In this case the required toolface heading ($\widehat{APY}$) is 75° right of High Side in order to change the azimuth to 060°. Notice also that this will result in a slight increase in inclination (OY represents about $8\frac{1}{2}$°). If the orientation angle were chosen as $\widehat{APX}$, the result would be a direction of 060°, but a drop in angle to about $7\frac{1}{2}$°.

On the rig, the directional driller may use a "Ouija board" (Fig. 3.16). This involves a similar technique to that described for the Ragland diagram but is much quicker since no diagrams need be drawn. Alternatively a series of equations can be derived to calculate the necessary angles directly.

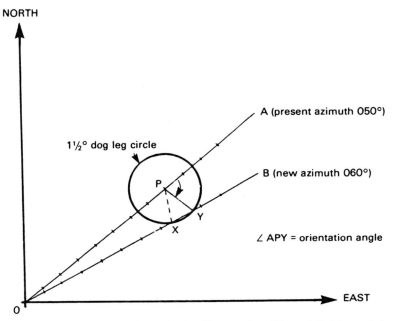

Fig. 3.15. Vector diagram to determine toolface setting. (Note: 1 division = 1 degree of inclination.)

To calculate the change in azimuth for a given toolface heading, dog-leg angle and inclination,

$$\Delta\beta = \tan^{-1}\left(\frac{\tan DL \sin TF}{\sin \alpha_1 + \tan DL \cos \alpha_1 \cos TF}\right)$$

where $\Delta\beta$ = change in azimuth
      DL = dog-leg angle
      TF = toolface setting
      α_1 = present inclination.

The new inclination angle can be given by

$$\alpha_2 = \cos^{-1}(\cos \alpha_1 \cos DL - \sin \alpha_1 \sin DL \cos TF)$$

To calculate the toolface heading from the dog-leg angle and the expected change in inclination, the following equation can be used:

$$TF = \cos^{-1}\left(\frac{\cos \alpha_1 \cos DL - \cos \alpha_2}{\sin \alpha_1 \sin DL}\right)$$

Orienting Procedure

Once the required toolface setting has been determined, the effect of reactive torque must be considered. Reactive torque is the twisting effect

34

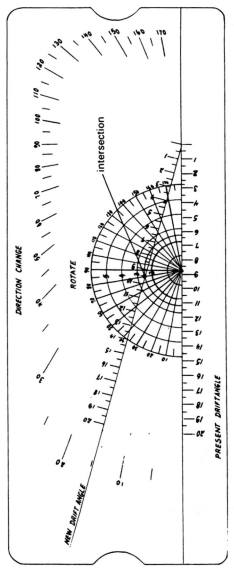

Fig. 3.16. Ouija board. For a present drift angle (inclination) of 8¾°, dog leg severity of 2¼° and direction change of 14° to the left. Intersection of new drift angle scale with dog leg circle gives the new inclination of 9½°, and toolface setting 80° left.

caused by the stator of the downhole motor turning anticlockwise in response to the rotor turning clockwise. The amount of twisting depends on the physical properties of the motor, the length of the drill string and the formation characteristics. Motor manufacturers produce estimates of how much left-hand turn can be expected under certain situations. From these tables, or from experience of drilling with similar tools in similar formations, the directional driller must compensate for reactive torque by deliberately pointing the tool face to the right of the calculated heading. As soon as the bit begins to drill, the scribe line will turn to the left to bring it back to the calculated heading. The amount of WOB will also affect the reactive torque. As the bit drills off, the reactive torque will reduce.

A series of surveys can be taken to ensure the scribe line is in fact aligned correctly. When a single shot surveying instrument is lowered into the hole on wireline, it will position itself within the orienting sub such that the reference mark is aligned with the toolface. When the survey picture is developed, the direction of the toolface can be read off directly (Fig. 3.17). If the toolface is not correct, the pipe must be turned at the rig floor. Any residual torque in the drill string should be removed by working the pipe. The orientation of the scribe line is then checked again by running another single shot. Several surveys may be necessary before the directional driller is satisfied that the toolface is correct. The use of a steering tool or MWD tool can reduce the amount of time taken to do this by sending continuous toolface results to surface. These continuous monitoring devices are also useful while the hole is being drilled, since the effect of reactive torque can

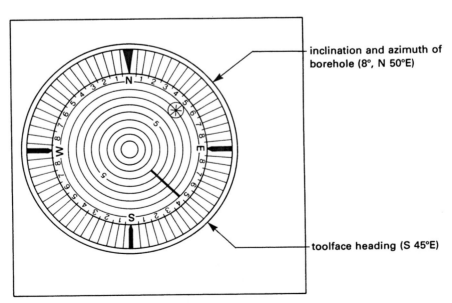

Fig. 3.17. Single shot picture showing orientation of scribe line.

be seen. If too much allowance was made for reactive torque in the initial orientation, the pipe can be turned at surface such that the correct heading is being shown on the surface readout.

SPECIALIZED DEFLECTION TECHNIQUES

In the planning of directional wells it is usually intended to drill vertically until the kick-off point (KOP) at which the well is deflected towards the target. When the reservoir is relatively shallow and covers a wide area, this method requires a very high angle of inclination to reach the periphery of the field. The angle could be reduced by selecting a shallower KOP, but this is not always possible, because of softer formations nearer the surface.

An alternative approach is to eliminate the vertical part of the trajectory completely by introducing deflection at the surface. By doing this, the problems of building and maintaining inclination are reduced, and a larger area of the field can be covered from one central platform. There are two techniques that have proved successful.

Curved Conductors

In offshore areas such as the Gulf of Mexico there are oil and gas reservoirs situated at depths of 3000–6000 ft below the seabed. Large fixed platforms are required to develop these fields in water depths up to 450 ft. Conventional directional wells from these platforms cannot, however, cover the

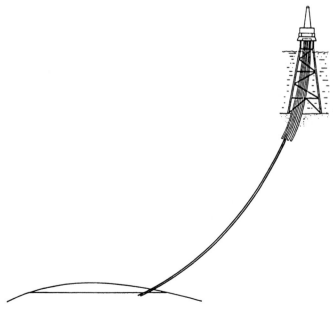

Fig. 3.18. Fixed platform using curved conductors.

entire area and so several platforms are required, increasing the cost of the development by a substantial amount.

It has been possible since the 1970s to install curved conductors on these platforms. The curvature of the conductors is 3–6° per 100 ft and they are driven 150–200 ft into the seabed. The conductors can be oriented towards the target as they are being driven, so that horizontal corrections for azimuth are reduced. The initial deflection at the seabed is already 10–20° of inclination. Drilling problems with the curved conductors have proved to be no more serious than in previous directional wells. The typical profile of such a well is shown in Fig. 3.18.

Slant Hole Drilling

Another technique for drilling an inclined hole starting from the surface is to mount the rig at some angle to the vertical. The hole is spudded at this angle and drilling continues in the conventional manner. It is theoretically possible to orient the rig such that the hole is drilled directly to the target. For some land-based operations, the derrick has been tilted at 45° to the vertical. An example of a slant hole profile is given in Fig. 3.19.

To adapt a standard drilling rig for slant hole drilling, certain modifications must be made:

(a) The rotary table must be inclined so that it is perpendicular to the axis of the derrick. (A power swivel may be used as an alternative.)

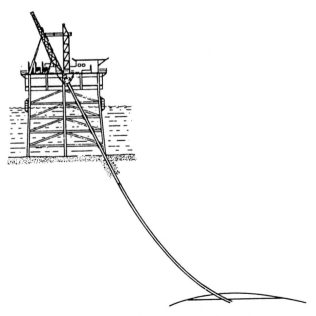

Fig. 3.19. Fixed platform using slant rig.

(b) The travelling block and hook must be run on guide rails up inside the derrick.

(c) The BOP stack must be mounted on a frame that can be tilted to the required angle.

(d) A hydraulic pipe racking system makes the handling of drill pipe much easier.

A slant hole drilling programme is being used on the Morecambe Bay gas field. The reservoir is located some 3000 ft below the seabed and has an area of 35 square miles. By conventional directional drilling, the wells could only achieve a maximum reach of about 3000 ft. With the derrick inclined at a maximum angle of 30°, a reach of almost 5000 ft was possible, increasing the area covered by a factor of almost 3.

QUESTIONS

3.1. To accommodate an MWD tool an 8-in. OD non-magnetic drill collar has to be bored out from 3-in. ID to 3.25-in. ID. By how much will this reduce the stiffness of the collar?

3.2. In a deviated well the proposed BHA calls for 8″ × 3″ drill collars run beneath 5″, 19.5 lb/ft drill pipe. Calculate the effects of bending for this configuration, and suggest an alternative to reduce bending.

3.3. A pendulum assembly is made up of a number of 8-in. drill collars whose weight is 147 lb/ft. The inclination of the hole is 20° and the tangent length is 40 ft. Calculate the side force on the bit.

3.4. Explain how the tendency for a bottom hole assembly to build or drop angle can be affected by changing the position of the lowermost stabilizer.

3.5. A correction run is to be made using a positive displacement motor and a bent sub that is expected to make a total change in angle of 2.5° per 100 ft drilled. The present inclination and azimuth are 25° and 160° respectively. If the toolface is set at 60° to the right of High Side, calculate the change in azimuth and the new inclination angle after drilling 100 ft with this assembly.

3.6. During a kick-off using jet deflection, a maximum dog-leg severity of 5° per 100 ft is permitted. The present inclination is 8° and the hole must be turned 25° to the left. If the toolface is set at 50° to the left of High Side:
    (a) what is the expected length of the correction run?
    (b) what is the final inclination?

3.7. To correct the path of the wellbore the angle must drop from 18° to 15° and the direction must turn to the left from its present heading of N 70° E. This must be accomplished over the next 400 ft, and the dog-leg severity is limited to 2° per 100 ft. Calculate the required toolface setting, allowing 30° reactive torque due to the motor. Calculate the new direction at the end of the correction run. Check your results by drawing a Ragland diagram.

3.8. At an inclination of 8°, the well must be turned to the right by setting a whipstock. The expected dog-leg angle is 3°. Draw a Ragland diagram to calculate

(a) the toolface which will give the maximum turn to the right,

(b) the resulting change in azimuth and direction.

Check your results by applying the appropriate equations.

FURTHER READING

"New components help make better drilling assemblies", W. D. Moore, *Oil and Gas Journal*, 5 March 1979.

"Jet-bit deflection saves money in directional wells", *Oil and Gas Journal*, 1 June 1959.

"Taking the kinks out of hole deflection", R. E. Brumley and J. McCollum, *Drilling*, June 1968.

"Curved well conductors and offshore platform hydrocarbon development", B. E. Cox and W. A. Bruha, O.T.C. paper no. 2621.

"Canadian operator succeeds in slant hole drilling project", B. M. Lowen and G. D. Gradeen, P.E.I., August 1982.

"Slant drilling holds the key to economics", *The Oilman*, March 1984.

Drilling Data Book, I.F.P. Editions Technip.

"Deviation tool controls bit walk", D. Keene and D. McKenzie, *Oil and Gas Journal*, 3 September 1979.

"Evaluating and planning directional wells utilizing post analysis techniques and a three dimensional bottom hole assembly program", K. K. Millheim, F. H. Gubler and H. B. Zaremba, S.P.E. paper no. 8339.

"Prediction of Drilling Trajectory in Directional Wells via a New Rock-bit Interaction Model", H-S Ho. S.P.E. paper no. 16658.

Chapter 4

DIRECTIONAL WELL PLANNING

Careful planning before the well is spudded can lead to substantial savings in the cost of drilling a directional well. Many factors influence the trajectory of the borehole. Some of these may be difficult to estimate (for example, the amount of bit walk that may occur in certain formations). The experience gained from drilling previous directional wells in the same area is therefore very useful and should be incorporated at the planning stage of the next well.

GENERAL CONSIDERATIONS

Reference Points and Coordinates

It is not uncommon for directional wells drilled from one platform to cover a reservoir several square miles in area. In planning wellpaths over such large distances some attention should be paid to the coordinate system adopted.

The most common method of fixing the position of a point on the Earth's surface is to give its latitude and longitude (Fig. 4.1). A line of latitude runs parallel to the equator, and is denoted by a number of degrees (0–90°) North or South of the equator. A line of longitude is perpendicular to the equator and passes through the North and South poles and is denoted by a number of degrees (0–180°) East or West of Greenwich. However, for the purposes of planning a directional well, it is more convenient if the curved surface of the Earth is projected onto a flat surface on which maps can be drawn. One such system is known as the Universal Transverse of Mercator (UTM)*. This is basically a projection of the section of the Earth's surface that contains the area of interest. In carrying out the projection there is some distortion of the axes such that UTM North is slightly offset from Geographic (True) North. This small difference is significant over large distances and so must be taken into account when converting coordinates from one system to another.

*(see Appendix, page 62).

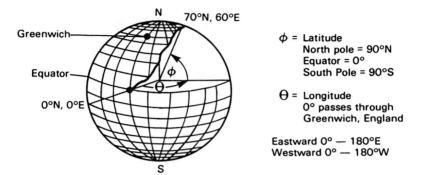

ϕ = Latitude
North pole = 90°N
Equator = 0°
South Pole = 90°S

Θ = Longitude
0° passes through
Greenwich, England

Eastward 0° — 180°E
Westward 0° — 180°W

Fig. 4.1. Latitude and longitude.

For the purposes of planning and monitoring, all measurements must be tied back to a common reference point. On offshore platforms this point is usually chosen to be the centre of the platform. All depths are measured from the elevation of the rotary table. If the target coordinates are given in the UTM system they must be converted and referenced back to the platform centre. All calculations are then simplified by adopting Northing and Easting coordinates. If the surface (slot) coordinate and the target

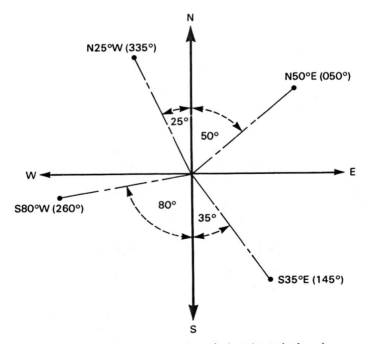

Fig. 4.2. Compass bearings (azimuth equivalents).

coordinates are known, the two end points of the trajectory are fixed. The target bearing (the direction in which the well must be drilled) can then be calculated. All directions must be given relative to True North before they can be used in survey calculations. Directions can be stated in two ways:

(i) quadrant system, in which the angle is measured 0–90° from North or South. Some examples are given in Fig. 4.2.
(ii) azimuth, in which the angle is measured 0–359° clockwise from North.

Target Zone

As well as defining a particular point as the target, the geologist will also specify a circular or rectangular area around that point known as the target zone. This allows the directional driller some tolerance on the final position of the well. A radius of 100 ft is commonly used as a target zone, but this will depend on particular requirements (for example, a relief well requires a much smaller target in order to be effective). The smaller the target zone the greater the number of correction runs necessary to ensure the target is intersected. This will result in longer drilling times and higher drilling costs. The target zone should therefore be as large as the geologist or the reservoir engineer can allow. The directional driller's job is then to place the wellbore within the target zone at minimum cost.

Formation Characteristics

The type of formations to be drilled can affect the planning of the profile in several ways.

(a) In selecting the kick-off point (KOP) the hardness of the formation is important. Hard formations may give a poor response to the deflecting tool, so that the kick-off may take a long time and require several bits. Kicking-off in very soft formations may result in large washouts. A soft–medium formation provides a better opportunity for a successful kick-off.
(b) Certain formations exhibit a tendency to deflect the bit either to the left or to the right. The directional driller can compensate for this effect by allowing some "lead angle" when orienting the deflecting tool. If the bit is expected to walk to the right by a certain number of degrees, the lead angle will point the bit an equal number of degrees to the left (Fig. 4.3). As the bit begins to drill, the formation effect will bring the well back on to its intended course.

Deflecting Tools Available

The capabilities of the deflecting tools available and the techniques that are applicable in a particular situation will influence the shape of the wellpath. If jet deflection is to be used, the KOP must be at a relatively shallow depth in a fairly soft formation. The availability of different bent subs will dictate the rate of build up. If a turbodrill is to be used over the long tangential section, it will tend to make the bit walk to the left. The directional

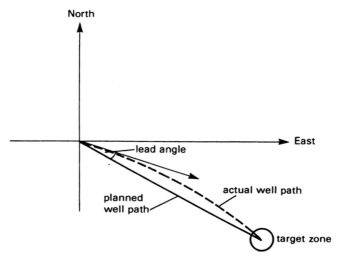

Fig. 4.3. Use of lead angle to compensate for bit walk.

behaviour of the various tools and bottom hole assemblies to be used must be considered when planning the well.

Location of Adjacent Wellbores

On offshore platforms there is only a small distance (7–12 ft) between adjacent conductors (Fig. 4.4). Under these conditions precise control is required and great care must be taken to avoid collisions directly beneath the platform. For this reason the KOPs for adjacent wells are chosen at varying depths to give some separation. When choosing slots it is better to allocate an outer slot to a target which requires large horizontal displacement. This will result in a shallower KOP to allow a smaller inclination. Slots closer to the centre of the platform should be allocated to targets requiring smaller inclinations and deeper KOPs (Fig. 4.5). This will help to avoid the problem of wells running across each other. As each well is being drilled, the proximity of all the adjacent wells should be checked by calculating inter-well distances from survey results. Anti-collision plots generated by computer are now widely available for doing this. It may be necessary to nudge a new well away from the existing wells, even though this means going away from the target direction. Once the well is a safe distance away, the well path can be corrected to bring it back onto the planned course (Fig. 4.6.).

Choice of Build-up Rate

If the change of angle occurs too quickly, severe dog-legs can occur in the trajectory. These sharp bends make it difficult for drilling assemblies and

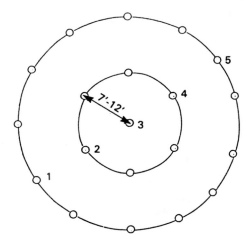

Fig. 4.4. Drilling slots in circular pattern.

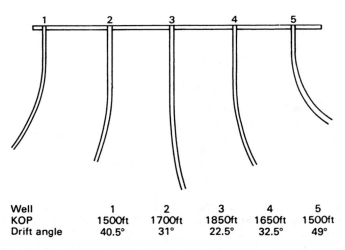

Well	1	2	3	4	5
KOP	1500ft	1700ft	1850ft	1650ft	1500ft
Drift angle	40.5°	31°	22.5°	32.5°	49°

Fig. 4.5. Slot allocation and choice of KOPs.

tubulars to pass through. Severe dog-legs also cause more wear on the drill string. If the angle is built up very slowly then it will take a longer interval of hole to reach the required inclination. To obtain a gradual build-up of angle at a reasonable curvature, a build-up rate of 1.5–2.5° per 100 ft is commonly used, but higher build-up rates may be necessary in some cases.

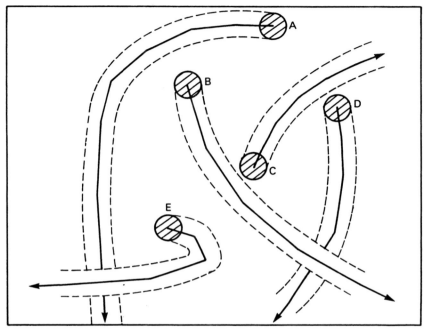

Fig. 4.6. Nudging well A away from adjacent conductors.

TYPES OF PROFILE

The well path may follow a number of different routes. The main types are summarized in the following paragraphs.

Type 1 (Build and Hold)

This is the most common and the simplest profile for a directional well. The hole is drilled vertically down to the KOP, where the well is deviated to the required inclination. This inclination is maintained over the tangential section to intersect the target (Fig. 4.7a). Generally, a shallow KOP is selected since this reduces the size of the inclination angle necessary to hit the target. This type of profile is often applied when a large horizontal displacement is required at relatively shallow target depths. Since there are no major changes in inclination or azimuth after the build-up section is complete, there are fewer directional problems with this profile. Under normal conditions the inclination should be 15–55°, although greater inclinations have been drilled.

Type II (Build, Hold and Drop)

This profile is similar to the Type I down to the lower part of the tangential section. Here the profile enters a drop-off section where the inclination is reduced, and in some cases becomes vertical as it reaches the target (Fig.

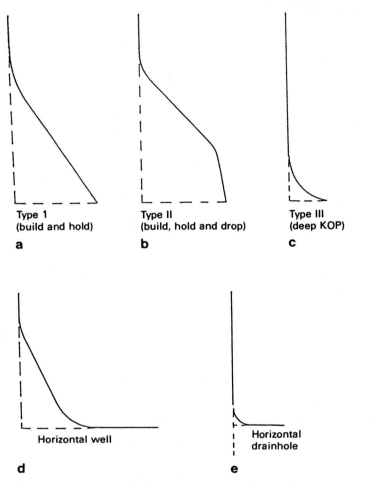

Type 1
(build and hold)

a

Type II
(build, hold and drop)

b

Type III
(deep KOP)

c

Horizontal well

d

Horizontal
drainhole

e

Fig. 4.7. Examples of directional well profiles.

4.7b). This is a more difficult profile to drill than the Type I, owing to the problems of controlling the drop-off section just above the target. Extra torque and drag can also be expected owing to the additional bend. This type of profile is used when the target is deep but the horizontal displacement is relatively small. (Under such conditions a Type I profile may produce a small inclination angle which would be difficult to control.) It also has applications when completing a well that intersects multiple producing zones, or in relief well drilling if it is necessary to run parallel with the wild well.

Type III (Deep Kick-off and Build)

This profile is only used in particular situations such as salt dome drilling or sidetracking (Fig. 4.7c). A deep KOP has certain disadvantages.

(a) Formations will probably be harder and less responsive to deflection.
(b) More tripping time is to change out BHAs while deflecting.
(c) Build up rate is more difficult to control.

Horizontal Wells

A horizontal well is one in which the inclination reaches 90° through the reservoir section. Horizontal wells have important applications in improving production from certain reservoirs that would otherwise be uneconomic (e.g. fractured limestone, low-permeability zones, etc.) The profile of the horizontal well is shown in Fig. 4.7d. Notice that there is more than one build-up section used to achieve the inclination of 90°. Conventional techniques are employed to drill this type of horizontal well, but there are many drilling problems to be overcome and so drilling costs are higher.

Horizontal Drainholes

In this type of profile the well is drilled vertically to the KOP using conventional techniques. A special BHA is then run which is used to build up angle rapidly along a circular arc of about 30 ft radius (Fig. 4.7e). This corresponds to build-up rate of 2° per foot. This rapid build-up of angle is only possible using special components in the drill string, such as articulated collars and knuckle joints. This type of profile can be used for producing from tight formations and reducing gas or water coning problems.

GEOMETRICAL PLANNING

Geometrical Planning for Type I Profile

The following information is required:

(a) surface (slot) coordinates;
(b) target coordinates;
(c) true vertical depth of target;
(d) true vertical depth to KOP;
(e) build-up rate.

The choice of slot depends on a number of factors including target location and the proximity of other wells. The target coordinates and depth are selected by the geologist. The choice of KOP and build-up rate has to be made by the directional planning engineer.

   The profile of the well is shown in Fig. 4.8. The coordinates of the points A, B, C and T must be determined on both horizontal and vertical views of the wellbore. Point A is defined by the surface coordinates, and point T by the target location. On the horizontal plan the displacement of the target (H_t) can be calculated by:

$$H_t = [(N_t - N_a)^2 + (E_t - E_a)^2]^{1/2}$$

where N_t = Northing of target
E_t = Easting of target
N_a = Northing of slot
E_a = Easting of slot

Notice that all these Northings and Eastings must be given with respect to the platform centre, or other reference point. The depths will be referenced back to the rotary table.

The proposed direction or target bearing β can also be calculated from the horizontal plan:

$$\beta = \tan^{-1}\left(\frac{E_t - E_a}{N_t - N_a}\right)$$

Knowing the displacement H_t and the depth of KOP (V_b), only the position of point C remains to be found. Point C is at the end of the build-up section when the maximum inclination is reached. In order to find the coordinates of C, the maximum angle of inclination must be determined. Let the build-up rate = ϕ degrees per 100 ft and let R = radius of curvature. By proportion

$$\frac{\phi}{360} = \frac{100}{2\pi R}$$

$$\Leftrightarrow R = \frac{18,000}{\pi\phi}$$

The inclination angle α is equal to the sum of angles x and y, which can be calculated from

$$\tan x = \frac{PF}{FT} = \frac{H_t - R}{V_t - V_b}$$

and

$$\sin y = \frac{PC}{PT}$$

where FT/PT = $\cos x$ and PC = R.

The angle α can therefore be determined from

$$\alpha = \tan^{-1}\left(\frac{H_t - R}{V_t - V_b}\right) + \sin^{-1}\left(\frac{R\cos x}{V_t - V_b}\right)$$

At point C:

$$BE = R\sin\alpha \qquad EC = R - R\cos\alpha = R(1 - \cos\alpha)$$

The arc BC can be determined from

$$\frac{BC}{2\pi R} = \frac{\alpha}{360} \Leftrightarrow BC = 100\frac{\alpha}{\phi}.$$

This allows the coordinates of C to be determined in Fig. 4.8:

True vertical depth $= V_c$ $= V_b + R \sin \alpha$
Horizontal departure $= H_c$ $= R(1 - \cos \alpha)$
Measured depth $= MD_c = MD_b + 100\alpha/\phi$

The measured depth at T can also be found:

$$MD_t = MD_c + CT$$

$$= MD_c + \frac{V_t - V_c}{\cos \alpha}$$

Note that the measured depth (MD) is the cumulative along-hole depth for that point from the reference point at surface.

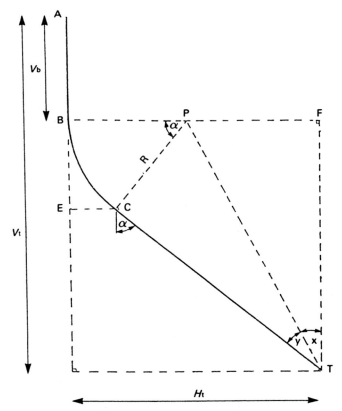

Fig. 4.8. Geometrical planning for Type I profile.

EXAMPLE 4.1

Using the following information, calculate the coordinates of a Type I well profile.

Slot coordinates	15.32 ft N, 5.06 ft E
Target coordinates	1650 ft N, 4510 ft E
TVD of target	9880 ft
KOP	1650 ft
Build up rate	1.5° per 100 ft

The horizontal displacement H_t is

$$H_t = [(1650 - 15.32)^2 + (4510 - 5.06)^2]^{1/2} = 4792.35 \text{ ft}$$

The target bearing β is

$$\beta = \tan^{-1}\left(\frac{4510 - 5.06}{1650 - 15.32}\right) = 70.1° \qquad \text{or N 70° E}$$

Radius of curvature R is

$$R = \frac{18,000}{\pi \times 1.5} = 3819.72 \text{ ft}$$

Maximum angle of inclination α is

$$\alpha = x + y$$

where

$$x = \tan^{-1}\left(\frac{4792.35 - 3819.72}{9880 - 1650}\right) = 6.74°$$

and

$$y = \sin^{-1}\left(\frac{3819.72 \times \cos 6.74°}{9880 - 1650}\right) = 27.45°$$

And therefore

$$\alpha = 6.74° + 27.45° = 34.19°$$

Coordinates of C are

$$V_c = 1650 + 3819.72 \sin 34.19 = 3796.45 \text{ ft}$$
$$H_c = 3819.72(1 - \cos 34.19°) = 660.13 \text{ ft}$$
$$MD_c = 1650 + \frac{100 \times 34.19}{1.5} = 3929.33 \text{ ft}$$

And at the target:

$$MD_t = 3929.33 + \frac{(9880 - 3796.45)}{\cos 34.19°} = 11,283.91 \text{ ft}$$

The profile for this well is shown in Fig. 4.9. Computer programs are now used to carry out these calculations and produce large-scale plots. The coordinates at regular intervals can be printed out along the wellbore.

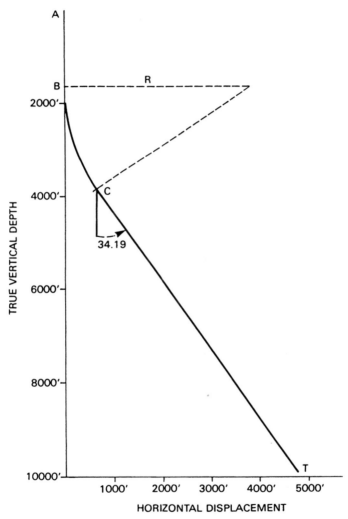

Fig. 4.9. Final profile for Example 4.1 (Type I).

Point	TVD (ft)	Horizontal displacement (ft)	MD (ft)
A	0	0	0
B	1650	0	1650
C	3796.45	660.13	3929.33
T	9880	4792.35	11,283.91

Geometrical Planning for Type II Profile

The following information is required:

(a) surface coordinates;
(b) target coordinates;
(c) true vertical depth of target;
(d) true vertical depth of KOP;
(e) rate of build-up;
(f) rate of drop-off;
(g) required TVD at end of drop-off;
(h) final angle of inclination through target.

On Figure 4.10 the distances V_b, V_e and V_t are known, as is the horizontal displacement H_t (determined from surface and target coordinates as before). The radius of curvature R_1 can be calculated from ϕ_1 (build-up

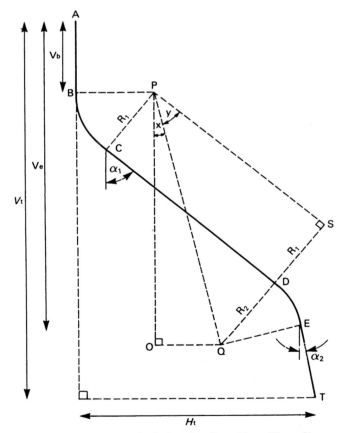

Fig. 4.10. Geometrical planning for a Type II profile.

rate). Likewise R_2 can be found from ϕ_2 (drop-off rate). The final inclination α_2 is known, but the inclination over the tangential section, α_1, must be calculated. Since PS is parallel with CD, and OP is vertical:

$$\text{angle } \alpha_1 = x + y$$

where

$$\tan x = \frac{OQ}{OP} \quad \text{and} \quad \tan y = \frac{QS}{PS}$$

$$OQ = H_t - R_1 - R_2 \cos \alpha_2 - (V_t - V_e) \tan \alpha_2$$
$$OP = V_e - V_b + R_2 \sin \alpha_2$$
$$QS = R_1 + R_2$$
$$PS = (PQ^2 - QS^2)^{1/2} \quad \text{where } PQ = (OP^2 + OQ^2)^{1/2}$$

Having calculated the four distances OQ, OP, QS and PS, the angles x and y can be determined and hence the unknown α_1 can be found.

The coordinates of the various points C, D, E and T can be determined as follows:

At point C
$$V_c = V_b + R_1 \sin \alpha_1$$
$$H_c = R_1 - R_1 \cos \alpha_1$$
$$MD_c = MD_b + \frac{100\alpha_1}{\phi_1}$$

At point D
$$V_d = V_c + PS \cos \alpha_1$$
$$H_d = H_c + PS \sin \alpha_1$$
$$MD_d = MD_c + PS$$

At point E
$$V_e \text{ (known)}$$
$$H_e = H_d + R_2(\cos \alpha_2 - \cos \alpha_1)$$
$$MD_e = MD_d + \frac{100(\alpha_1 - \alpha_2)}{\phi_2}$$

At point T
$$MD_t = MD_e + \frac{V_t - V_e}{\cos \alpha_2}$$

EXAMPLE 4.2

Using the following information, determine the trajectory of a Type II profile which will intersect the target.

Horizontal displacement (H_t)	= 6000 ft
Target depth (V_t)	= 12,000 ft
TVD of KOP (V_b)	= 1500 ft
Build-up rate (ϕ_1)	= 2° per 100 ft
Drop-off rate (ϕ_2)	= 1.5° per 100 ft
TVD at end of drop (V_e)	= 11,000 ft
Final inclination (α_2)	= 20°

$$R_1 = \frac{18,000}{\pi \times 2} = 2864.79 \text{ ft} \quad \text{and} \quad R_2 = \frac{18,000}{\pi \times 1.5} = 3819.72 \text{ ft}$$

$$\begin{aligned}
OQ &= 6000 - 2864.79 - (3819.72 \cos 20°) \\
&\quad - (12,000 - 11,000) \tan 20° \\
&= -818.12 \text{ ft}
\end{aligned}$$

$$\begin{aligned}
OP &= 11,000 - 1500 + 3819.72 \sin 20° \\
&= 10,806.42 \text{ ft}
\end{aligned}$$

$$\begin{aligned}
QS &= 2864.79 + 3819.72 \\
&= 6684.51 \text{ ft}
\end{aligned}$$

$$\begin{aligned}
PQ^2 &= (10,806.42)^2 + (-818.12)^2 \\
\Leftrightarrow PQ &= 10,837.34 \text{ ft}
\end{aligned}$$

$$\begin{aligned}
PS^2 &= (10,837.34)^2 - (6684.15)^2 \\
\Leftrightarrow PS &= 8530.26 \text{ ft}
\end{aligned}$$

$$\text{Angle } x = \tan^{-1}\left(\frac{-818.12}{10,806.42}\right) = -4.33°$$

$$\text{Angle } y = \tan^{-1}\left(\frac{6684.51}{8530.26}\right) = 38.08°$$

Therefore inclination $\alpha_1 = -4.33° + 38.08° = 33.75°$.

Note that the negative value for angle x is due to the fact that the geometry is slightly different from Figure 4.10 in that point Q lies to the left of P. The calculation method, however, remains valid. The coordinates therefore are as follows:

$$\begin{aligned}
V_c &= 1500 + (2864.79 \sin 33.75°) = 3091.59 \text{ ft} \\
H_c &= 2864.79(1 - \cos 33.75°) = 482.80 \text{ ft} \\
MD_c &= 1500 + (100 \times 33.75)/2 = 3187.50 \text{ ft} \\
V_d &= 3091.59 + (8530.26 \cos 33.75°) = 10,184.24 \text{ ft} \\
H_d &= 482.80 + (8530.26 \sin 33.75°) = 5221.96 \text{ ft} \\
MD_d &= 3187.50 + 8530.26 = 11,717.76 \text{ ft} \\
V_e &= 11,000 \text{ ft} \\
H_e &= 6000 - 3819.72(\cos 20° - \cos 33.75°) = 5586.62 \text{ ft} \\
MD_e &= 11,717.76 + 100(33.75° - 20°)/1.5 = 12,634.43 \text{ ft} \\
V_t &= 12,000 \text{ ft} \\
H_t &= 6000 \text{ ft} \\
MD_t &= 12,634.43 + (12,000 - 11,000)/\cos 20° = 13,698.61 \text{ ft}
\end{aligned}$$

The final trajectory is shown in Fig. 4.11. Note that these equations can also be used for the conventional S shape well for which the inclination $\alpha_2 = 0°$ and $V_e = V_t$.

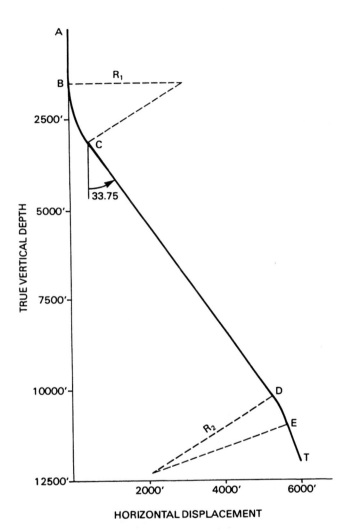

Fig. 4.11. Final profile for Example 4.2 (Type II).

Point	TVD (ft)	Horizontal displacement (ft)	MD (ft)
A	0	0	0
B	1500	0	1500
C	3091.59	482.80	3187.50
D	10,184.24	5221.96	11,717.76
E	11,000	5586.62	12,634.43
T	12,000	6000	13,698.61

Geometrical Planning for Type III Profile

The following information is required:

(a) surface coordinates;
(b) target coordinates;
(c) one further parameter from
     (i) vertical depth at KOP,
     (ii) build up rate,
     (iii) maximum angle of inclination.

If any one of the parameters (i), (ii) or (iii) is known, the others can be determined.

Knowing V_t, H_t *and KOP depth* (V_b). From Fig. 4.12 it can be seen that

$$\frac{\alpha_1 + \alpha_2}{2} = \tan^{-1}\left(\frac{H_t}{V_t - V_b}\right)$$

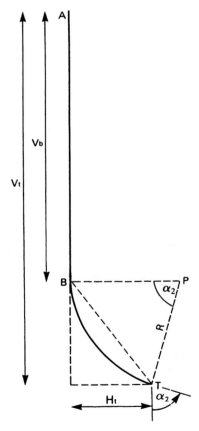

Fig. 4.12. Geometrical planning for a Type III profile.

and since $\alpha_1 = 0$, then $\alpha_2 = 2 \tan^{-1}[H_t/(V_t - V_b)]$. Also

$$\sin \alpha_2 = \frac{V_t - V_b}{R} \Leftrightarrow R = \frac{V_t - V_b}{\sin \alpha_2}$$

$$\frac{\text{arc length BT}}{2\pi R} = \frac{\alpha_2}{360}$$

$$\Leftrightarrow BT = \frac{2\pi(V_t - V_b)}{\sin \alpha_2} \frac{\alpha_2}{360}$$

$$\text{Build up rate } \phi = \frac{\alpha_2}{BT} = \frac{180 \, (\sin \alpha_2) \times 100}{\pi(V_t - V_b)}$$

$$\text{or } \phi = \frac{18,000}{\pi R} \text{ (degrees per 100 ft)}$$

EXAMPLE 4.3

Given the following information, calculate the trajectory of a Type III well.

Horizontal displacement (H_t) = 1500 ft
TVD to target (V_t) = 10,000 ft
Depth of KOP (V_b) = 7000 ft

$$\text{Final inclination } \alpha_2 = 2 \tan^{-1}\left(\frac{1500}{10,000 - 7000}\right) = 53.13°$$

$$R = \frac{10,000 - 7000}{\sin 53.13°} = 3750.00 \text{ ft}$$

$$\text{Build-up rate } (\phi) = \frac{18,000}{\pi \times 3750}$$
$$= 1.53° \text{ per 100 ft}$$

$$MD_t = 7000 + (53.13/1.53)100 = 10,472.55 \text{ ft}$$

This trajectory is shown in Fig. 4.13.

COMPUTER APPLICATIONS

In the previous section, the method of calculating the coordinates of the three basic directional profiles was described. Calculating the position of each point by hand is a fairly lengthy procedure, although the calculations themselves are not difficult. It must be remembered, however, that only turning in the vertical plane has been considered. Allowing for any expected tendency for the bit to walk to the left or right requires some further calculations in the horizontal plane. In reality a number of different trajectories may have to be investigated in order to reduce potential

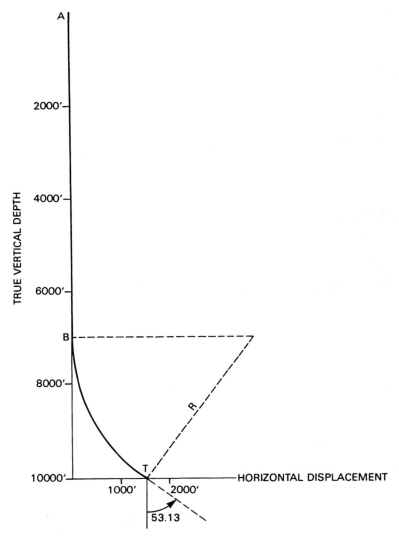

Fig. 4.13. Final profile for Example 4.3 (Type III).

Point	TVD (ft)	Horizontal displacement (ft)	MD (ft)
A	0	0	0
B	7000	0	7000
T	10,000	1500	10,472.55

drilling problems. These problems may include excessive torque and drag, differential sticking and borehole instability. Each choice of KOP or build-up rate will require a new trajectory to be calculated and plotted. Such repetitive calculations are best handled by computers, which are becoming more common in drilling operations.

When planning wells from large offshore platforms, such computer techniques are essential to handle the large amount of data and plan the wells efficiently. Many computer packages are available from directional drilling service companies and several operating companies have their own in-house programs. The programs take much of the labour out of planning and save a lot of time. The computer package can offer many useful features.

(a) The computer package can generate well plans quickly and accurately for various combinations of kick-off point, build-up rates, drop-off rates, etc. Walk rates to the left or right over certain intervals can also be incorporated easily into the trajectory. The actual coordinates along the planned trajectory can be calculated at regular intervals (say, every 50 ft). The computer can also produce large-scale plots of each trajectory that can be used to compare the various options. Any potential problems with a particular profile can be discussed between the operating company and the directional drilling service company prior to spudding the well.

(b) The computer has a large memory capacity that can be used to store the survey data from existing wells already drilled from the platform. The planned trajectory of future wells can also be stored in the computer. By accessing the relevant data files, the current well being planned can be viewed in relation to these other adjacent wells. Plots showing one well superimposed upon another in both vertical and horizontal planes can be generated. The horizontal structure plot shows a plan view of all the wells drilled from the platform, with the actual measured depth or true vertical depth marked at regular intervals along the well path (Fig. 4.14).

(c) As the well is being drilled, the survey data from the well site can be analysed by the computer, which will then calculate the actual trajectory and compare it with the planned trajectory. The dog-leg severity can also be calculated to assess the sharpness of bends in the trajectory.

(d) Having calculated the actual position of the well being drilled, the computer can also find the distance and direction from that point to any other well in close proximity. This information can be plotted to give a visual representation of how close adjacent wellbores are to the current well (Fig. 4.15). These plots can be used either at the planning stage or while the well is being drilled, to avoid possible intersections. The plots can be updated rapidly as more information becomes available.

(e) Despite the improvements in the accuracy of surveying instruments, the position of any point along the wellbore is subject to error. The

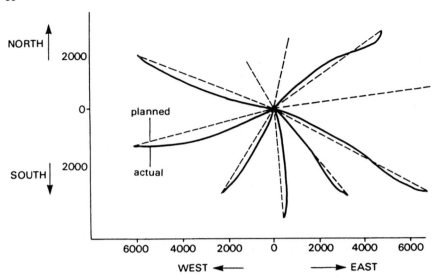

Fig. 4.14. Horizontal plan of wells drilled from one platform.

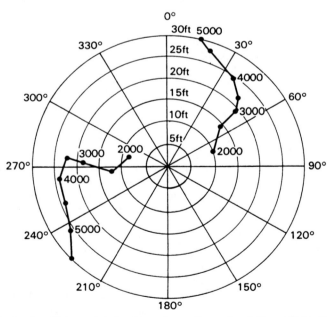

Fig. 4.15. Anti-collision plot, looking down the axis of the reference well. The distance between this reference well and other adjacent wells at particular depths can be found by interpolating between the contour lines.

amount of error associated with a particular point can be expressed in terms of an area of uncertainty around that point. The size of the area of uncertainty depends on the accuracy to which the depth can be measured, on instrument error, and on other factors. A model for assessing the accuracy of the borehole position can be incorporated into a computer program and used to plot the area or ellipse of uncertainty related to each survey point.

QUESTIONS

4.1. A directional well is to be drilled from an offshore platform to intersect a target whose horizontal displacement is 3500 ft at a depth of 10,500 ft (TVD). A Type I profile (build and hold) is to be used with a KOP = 1600 ft and a build-up rate of 1.5° per 100 ft. Calculate:
 (a) the inclination at the end of the build section;
 (b) the horizontal displacement and TVD at the end of the build section;
 (c) the total measured depth to the target.

4.2. A Type II profile (build, hold and drop) is planned for the following well: V_t = 10,000 ft, H_t = 4000 ft, build up and drop-off rate = 2° per 100 ft. At the end of the drop-off section the inclination is to be 0° at a TVD of 9000 ft. If the inclination over the tangential section is not to exceed 50°, what is the deepest kick-off point that will meet the requirements of the well?

4.3. The following parameters are to be used to plan an S shaped well: KOP = 1500 ft, V_t = 10,000 ft, H_t = 6250 ft, build-up rate = 3° per 100 ft, drop-off rate = 2° per 100 ft. The end of the drop off section is to intersect the target at an inclination of 0°. Calculate:
 (a) the inclination angle over the tangential section;
 (b) the horizontal displacement and TVD at the end of the build section and at the start of the drop section;
 (c) the total length of the well.

4.4. To reach a target located 8000 ft away at a depth of 7000 ft, the operator is considering the use of a slant rig, as opposed to a conventional Type I profile. Compare the maximum inclination and the total measured depth for each option.
 (a) For the Type I plan, assume a KOP = 1500 ft and a build rate of 2° per 100 ft.
 (b) For the slant rig, assume an initial inclination of 30°, a KOP of 1500 ft (MD) and a build rate of 2° per 100 ft.

4.5. A Type III profile is planned for the following well: V_t = 6000 ft; H_t = 1800 ft; build rate = 2.5° per 100 ft. Calculate the KOP, the final inclination and the total depth of the well.

FURTHER READING

"Planning of directionally drilled wells in the offshore Wilmington Field using the hand-held calculator", D. D. Clark and J. W. Barth, I.A.D.C./S.P.E. paper no. 11360.

"Computerised well planning for directional wells", H. Hodgson and S. G. Varnado, S.P.E. paper no. 12071.

"Planning the directional well—a calculation method", W. H. McMillan, *Journal of Petroleum Technology*, June 1981.

"North Sea gas area drilling improved", H. A. Kendall and T. A. Morgan, *Oil and Gas Journal*, 2 February 1976.

"Designing well paths to reduce torque and drag", M. C. Sheppard, C. Wick and T. Burgess, S.P.E. paper no. 15463.

"Coordinate Systems and Map Projections", D. H. Maring, George Philip & Son (1980).

"Transverse Mercator Projection-Constants, Formulae and Methods", *Ordnance Survey*, HMSO 1983.

APPENDIX

Universal Tranverse Mercator (UTM) Projection

This system is based on a globe being inserted within a cylinder, and touching the cylinder along a chosen reference meridian. The curved surface of the globe is then projected on to the cylinder which is unrolled to form a map. The surface of the Earth is divided into grid sectors, each sector covering 6° of longitude and 8° of latitude. The co-ordinates of each grid sector are measured from an origin which is located at the Equator, and a point 500 000 m west of the central reference meridian.

The scale of the grid sector is correct along the central meridian, but is increased on either side. To account for this distortion a scale factor is applied. For practical purposes the scale factor can be taken as constant over a 10 km square block.

Convergence is defined as the angle between UTM North and True North. Lines of longitude within a grid section will converge towards the central meridian. Convergence may be positive (to the East of True North) or negative (to the West of True North).

If the platform reference points 0 and target A are given in UTM co-ordinates (denoted by N, E) the local rectangular co-ordinates (N_a, E_a) can be calculated knowing the convergence c. By subtraction the distances ΔN, ΔE can be found:

$$\text{angle } \gamma = \tan^{-1}\left(\frac{\Delta E}{\Delta N}\right) \text{ and distance } 0A = \frac{\Delta E}{\sin \gamma}$$

Local co-ordinates $E_a = 0A \sin (\gamma + c)$

$$= \frac{\Delta E}{\sin \gamma} \{\sin \gamma \cos c + \cos \gamma \sin c\}$$

$$= \Delta E \cos c + \Delta N \sin c$$

Similarly, $N_a = 0A \cos (\gamma + c)$

$$= \frac{\Delta E}{\sin \gamma} \{\cos \gamma \cos c - \sin \gamma \sin c\}$$

$$= \Delta N \cos c - \Delta E \sin c$$

Chapter 5

POSITIVE DISPLACEMENT MOTORS

In the search for a viable alternative to conventional rotary drilling in which the entire drill string is rotated from surface, various types of downhole motors have been proposed. Drill string rotation can be eliminated by having a motor placed downhole to drive the bit by hydraulic or electrical power. The turbodrill, first patented in 1873, provided a means of driving the bit by harnessing the power of the drilling fluid, but the complexity of the tool and the lack of suitable drilling bits prevented it from being widely used except in the USSR. The electrodrill, dating from 1891, was another tool pioneered largely by the Russians. An electric cable was run down the drill string to power an electric motor mounted above the bit. The motor drove the bit through a gearing mechanism. Although the possibilities of using electrodrills are still being investigated, no commercial tool is presently in use.

After World War II, another type of downhole motor was developed by the Smith Tool Company in the USA. This type of motor became known as the "positive displacement motor" or PDM. The design is based on the work done by René Moineau in developing Archimedian screw pumps. Archimedes invented a pump which could be used to lift water up from a lake. The pump consisted of a spiral shaft which could be turned by hand. The section end was lowered at an angle into the water and as the shaft was rotated water was discharged at the upper end. In the 1930s, Moineau developed this idea for pumping viscous fluids used in industrial processes. For use as a downhole motor, Moineau's pump could be operated in reverse so that mud pumped through the tool would cause the shaft to rotate and thus drive the bit.

The first commercial PDM was introduced by Smith in the late 1950s. It consisted basically of a steel shaft that rotated within a rubber-sleeved stator. As mud was pumped between the spiral shaft and the stator, the shaft rotated clockwise and turned the bit. The hydraulic power of the mud in terms of pressure and flow rate was converted to mechanical power at the bit in terms of rotational speed and torque. The PDM provided higher rpm at the bit than conventional rotary methods could achieve. The tool was first intended to compete with conventional rotary drilling for straight holes, but the short life of the bearings in the motor reduced its effectiveness. The rubber components in the tool were also sensitive to high temperatures.

In the 1960s, however, the PDM did find applications in directional drilling when used with a bent sub for kick-off operations. The bent sub placed above the motor could be oriented in the required direction. Since the PDM eliminated drill string rotation, the rotary table could be locked. The well was drilled by pumping mud through the motor, which rotated the bit and kicked the well off in a specific direction. This proved to be a quicker method than using a whipstock and gave a smooth curvature with less risk of severe dog-legs. Owing to the shorter length, the PDM was preferred to the turbodrill for this purpose.

In the USSR, development work began on multilobe motors. Instead of the PDM rotor having a circular cross-section, these new rotors had more complex geometries with up to nine lobes. The stator had one more lobe than the rotor, so that a fluid passageway was created. The advantage of these motors is that the rpm is reduced, though the torque is increased.

With deeper wells being drilled and more directional drilling, there was increased interest in PDMs during the 1970s. More service companies began manufacturing their own motors, notably Schlumberger, Baker and Christensen. In the 1980s, Russian PDMs were made available to the West through Drilex. Improvements in design and materials have enabled PDMs to run continuously for 100–200 hours. Both single and multilobe models are used for a wide range of applications including straight hole drilling, kick-offs, sidetracks, coring and geothermal wells.

MOTOR DESCRIPTION

The positive displacement motor consists of several components, shown in Fig. 5.1. These components are described in the following paragraphs.

Dump Valve

To prevent the motor rotating while running into the hole or pulling out of the hole, a by-pass valve or dump valve is installed at the upper end of the motor. This valve has radial ports that allow communication between the drill string and the annulus. During a trip when the pumps are shut off, these ports are open to allow the drill string to drain while pulling out or fill when running in. The ports must be closed off during drilling to allow normal flow through the motor. Under the increased pressure due to the mud pumps, a piston pushes a sleeve down to cover and seal off the ports. Whenever the pumps are shut off, a spring forces the sleeve upwards, opening the ports again. (Fig. 5.2).

Motor Section

The positive displacement motor consists of two basic components.

(a) The rotor is a steel shaft which is shaped in the form of a spiral or helix.

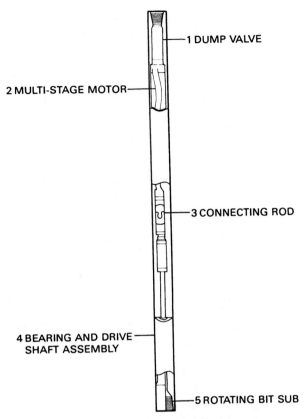

1 DUMP VALVE

2 MULTI-STAGE MOTOR

3 CONNECTING ROD

4 BEARING AND DRIVE
SHAFT ASSEMBLY

5 ROTATING BIT SUB

Fig. 5.1. Major components of a positive displacement motor (courtesy of Smith International Inc.).

For a single-lobe motor the cross-section of the shaft is circular. It is free at the top but attached to the universal joint at the bottom.

(b) The stator is a moulded rubber sleeve that forms a spiral passageway to accommodate the rotor. The rubber sleeve is fixed to the steel body of the motor.

When the rotor is fitted inside the stator, the difference in geometry between the two components creates a series of cavities. When the drilling fluid is pumped through the motor, it seeks a path between the rotor and stator. In doing so the mud displaces the shaft, forcing it to rotate clockwise as the mud continues to flow through the passageways. Positive displacement motors will operate with either gas or liquid as the drilling fluid.

In a single-lobe motor the flow area is relatively large, allowing a greater flow and hence a fast speed of rotation. In a multilobe motor there can be up to ten cavities in the stator and nine lobes on the rotor (Fig. 5.3). In this configuration the flow area is reduced and so the speed of rotation is also

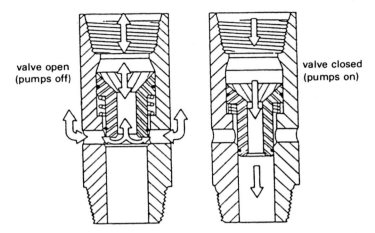

valve open
(pumps off)

valve closed
(pumps on)

Fig. 5.2. Operation of a by-pass valve (dump valve) (courtesy of Smith International Inc.).

reduced. The reduction in rpm and the corresponding increase in torque have applications in coring operations. The various configurations of positive displacement motors are usually denoted by the ratio of the number of lobes on the rotor to the number of lobes on the stator. This also relates to the rotor pitch and the stator pitch such that

$$i = \frac{C_r}{C_s} = \frac{P_r}{P_s}$$

where C_r = number of lobes on the rotor
C_s = number of lobes on the stator
P_r = pitch of the rotor
P_s = pitch of the stator.

single lobe (1/2)   multi-lobe (9/10)

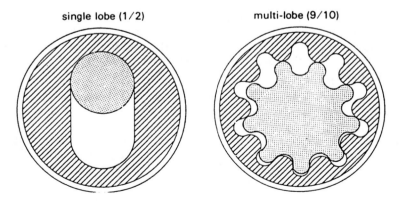

Fig. 5.3. Cross-sections through motors.

The first type of PDM was the single-lobe motor ($i = 1/2$), but 3/4 5/6 and 9/10 configurations are now also available. More stages or steps can be incorporated to increase the power and torque delivered by a PDM. One stage is equivalent to a complete spiral of the stator. There are usually three stages in the motor, but in some cases four or six stages are used. To achieve more power output, therefore, the overall length of the motor will increase.

Since the rotor is in contact with the stator as it rotates, special attention must be given to the materials used in the manufacture of each component. For efficient operation, a pressure-tight chamber must exist between the rotor and stator, but at the same time the shaft must rotate without causing excessive wear. The helical shaft is made of steel alloy and its surface is treated to increase its wear resistance. The choice of material for the stator is critical. Various rubber and elastomer materials have been tried and tested. Most of the elastomer components are susceptible to high temperatures. They are also affected by oil-based muds, which cause swelling. Polybutadiene has proved successful in durability tests but is difficult to mould into the required shape. With improved elastomer compounds PDMs can withstand temperatures up to around 350°F.

Universal Joint

Since the shaft is rotating eccentrically, the lower end must be connected to a universal joint. This joint converts the eccentric motion to concentric motion, which is then transmitted to the bit. Various types of flexible joints can be used, but the simplest design is a ball joint lubricated by grease. A rubber sleeve around the joint prevents contamination by mud. The universal joint is then connected to the drive shaft, which rotates within the bearing assembly. Two universal joints can be used, as shown in Fig. 5.4.

Bearing Assembly

This is probably the most critical component of the PDM, since the durability of the bearings very often determines the operating life of the motor itself. The bearing assembly fulfils two functions.

(a) It transmits the axial load to the bit. This is achieved through thrust bearings that consist of steel balls contained within spring-loaded ball races. Several sets of thrust bearings may be necessary to increase the load-carrying capacity of the motor.

(b) It maintains the central position of the drive shaft to ensure smooth rotation. This is done by using radial bearings. These are sleeve type bearings made of elastomer material. Two radial bearings are normally included in the assembly. The upper bearing acts as a flow restrictor, diverting a small percentage of mud through the bearing assembly for lubrication. Sealed bearings lubricated by oil will increase the operational life of the motor.

To measure the amount of bearing wear that has occurred during a motor run, a simple check can be made at the rotary table. While the PDM

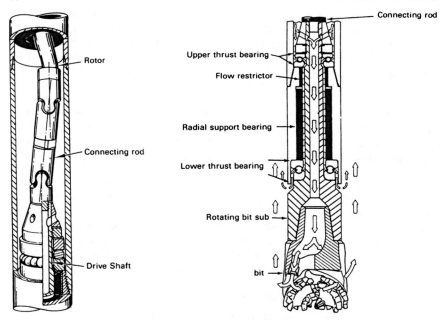

Fig. 5.4. Universal joint and bearing assembly (courtesy of Smith International Inc.).

is suspended and hanging freely, the distance between the bearing housing and the drive sub should be measured. When the motor is set down on the rig floor, the same measurement is taken again. The difference between these two measurements is referred to as the bearing wear. This can be compared to the maximum allowable bearing wear recommended by the motor manufacturer (Fig. 5.5).

PERFORMANCE CHARACTERISTICS

The energy supplied by the drilling fluid is used to rotate the helical shaft, which in turn rotates the bit. The only part of the motor that can be seen by an observer to rotate is the sub at the lower end. The pressure of the fluid trapped within the sealed chambers of the motor forces the shaft to turn. Only by displacing the shaft can the fluid exit from the passageway between the rotor and stator.

The speed of rotation n can be determined from

$$n = \frac{\text{flow rate}}{\text{specific displacement per revolution}}$$

$$\Leftrightarrow n = \frac{Q}{AP_rC_s}$$

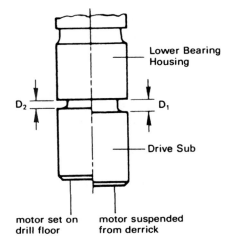

Tool Size O.D.	Max. tolerance $D_1 - D_2$	
inch	mm	inch
1 3/4	1.3	0.051
2 3/8	1.7	0.067
2 3/4	2.5	0.098
3 3/4	4.0	0.157
4 3/4	4.0	0.157
6 1/4	6.0	0.236
6 3/4	6.0	0.236
8	8.0	0.315
9 1/2	8.0	0.315
11 1/4	8.0	0.315

Fig. 5.5. Allowable bearing wear on motor (courtesy of Eastman-Christensen).

where Q = flow rate through the motor
A = cross-sectional area of the flow path
P_r = rotor pitch
C_s = number of lobes of the stator.

The cross-sectional area A of the flow path is a function of various geometrical factors, including the configuration of lobes, the rotor diameter and the eccentricity (Fig. 5.6). For a single-lobe motor the flow area is given by

$$A = \left(\frac{\pi d_r^2}{4} + 2ed_r\right) - \frac{\pi d_r^2}{4}$$

therefore

$$\Leftrightarrow A = 2ed_r$$

where d_r = diameter of rotor
        e = eccentricity.

The eccentricity is the distance between the centre of the rotor and the central axis of the stator (see Fig. 5.6). The speed of rotation for a single-lobe positive displacement motor is therefore given by

$$n = \frac{Q}{2ed_rP_rC_s}$$

In terms of oilfield units, this equation becomes

$$n = \frac{Q}{2ed_rP_rC_s} \times 12^3 \times 0.1337$$

$$\Leftrightarrow n = \frac{57.75Q}{ed_rP_r}$$

where Q = flow rate (gpm)
        e = eccentricity (in.)
        d_r = shaft diameter (in.)
        P_r = rotor pitch (in.)
        n = rotational speed (rpm).

It can be seen from the above equation that rotational speed increases linearly with flow rate. By increasing the geometrical factors e, d_r and P_r, the speed will be reduced. A range of different designs can therefore be produced to give varying bit speeds. These can be classified as low-speed motors (60–200 rpm), medium-speed motors (250–550 rpm) and high-speed motors (600–900 rpm).

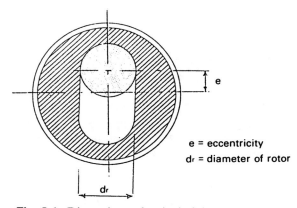

e = eccentricity
dr = diameter of rotor

Fig. 5.6. Dimensions of a single-lobe motor.

The power developed by the motor is calculated from the product of torque and angular velocity (in radians per second). In oilfield units this becomes

$$P = \frac{T}{550}\left(\frac{2\pi}{60}\right)n = 1.904 \times 10^{-4}Tn$$

where P = power (hp)
 T = torque (ft lb)
 n = motor speed (rpm)

The torque developed by the motor can be calculated by considering the relationship

$$P = \frac{Q\,\Delta p}{1714}$$

where Q = flow rate (gpm)
 Δp = pressure drop (psi)
 P = power (hp).

By combining these two equations an expression for torque can be derived:

$$1.904 \times 10^{-4}Tn = \frac{Q\Delta p}{1714}$$

where

$$Q = \frac{ned_rP_r}{57.75}$$

and solving for T yields

$$T = 5.306 \times 10^{-2} \times ed_rP_r\Delta p$$

This equation can therefore be used to calculate the theoretical torque delivered by a single-lobe PDM. Note that torque is directly proportional to the pressure drop through the motor but is independent of the rotational speed. The actual torque and power at the bit, however, will be less than that calculated, since the effect of losses has not yet been included. The efficiency of the motor must also be considered:

$$E = \frac{\text{useful power at the bit}}{\text{total power of the drilling fluid}}$$

This efficiency factor takes into account losses due to:

(a) internal fluid leaks along contact surfaces which will increase with wear (multilobe motors have lower efficiencies than single-lobe motors owing to there being more contact surfaces);
(b) friction losses in the bearings and universal joint;
(c) entry and exit losses.

The overall efficiency of a PDM depends on several factors and the value for a particular motor should be obtained from the manufacturer. For a

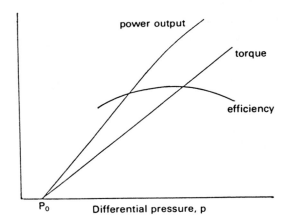

Fig. 5.7. Typical performance curves for PDMs.

single-lobe motor, E may vary between 70 and 90%. The efficiency tends to reduce with increasing number of lobes, so that for a multilobe motor the value of E may be 40–70%.

The performance characteristics for a particular PDM are usually given in the form of graphs. Examples of some typical curves are shown in Fig. 5.7. Comparison of different lobe profiles are given in Table 5.1.

APPLICATIONS OF PDMs

Although the positive displacement motor was first introduced for straight hole drilling, the major applications are now in directional wells.

TABLE 5.1 Comparison of Operating Characteristics for Different Motor Configurations

Rotor/stator configuration	1/2	3/4	5/6	9/10
Length (ft)	21	22.2	23.1	23.5
Weight (lb)	2350	2100	3290	3400
Maximum pressure drop across bit (psi)	500	1500	1000	1500
Flow rate (gpm)	325–450	250–500	350–600	200–650
Speed range (rpm)	230–332	65–135	90–160	55–185
Maximum pressure drop across motor (psi)	360	250	390	800–1000
Maximum torque (ft lb)	1160	2100	4200	4500–6000
Maximum power output at max. rpm and torque (hp)	73	54	128	211

Kick offs and correction runs
When used with a bent sub, PDMs are widely used as a deflecting tool. The bent sub can be oriented in the required direction and the drill string is not rotated. An MWD tool or steering tool will provide continuous monitoring of the toolface while drilling. Generally a low-speed, high-torque motor is best for this application. The length of the motor should be as short as possible to allow for the curvature of the wellpath.

Performance drilling
Improvements in the design of motors have made them more competitive with rotary methods. In particular, better bearing design has enabled

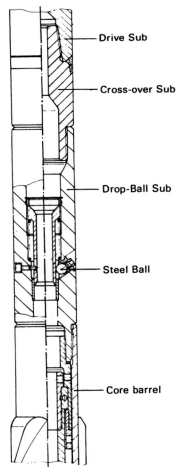

Drive Sub

Cross-over Sub

Drop-Ball Sub

Steel Ball

Core barrel

Fig. 5.8. Use of drop-ball sub when coring with a PDM (courtesy of Eastman-Christensen).

motors to operate downhole for longer periods, which improves the overall economics of the bit run. More powerful multistage motors have been introduced for drilling straight sections of hole. These tools, however, are also competing against turbodrills, which are generally more powerful and have higher rotational speeds than PDMs. As with turbodrills, the drill string is rotated for performance drilling to reduce the risk of stuck pipe.

Coring operations
High torque at low speeds makes a positive displacement motor suitable for coring. In a directional well, coring by rotary methods leads to casing and drill pipe wear owing to drill string rotation. Penetration rates are also low, making coring operations very expensive. With a multilobe PDM, less WOB is required when coring owing to the higher torque. Experience in the North Sea shows that good recovery rates and faster ROP are obtainable when coring with a PDM.

In conventional rotary coring, a steel ball is dropped from surface just before coring begins. Initially, mud is circulated through the inner barrel to flush out any debris. When the steel ball is dropped the flow of mud is diverted through the annulus between the inner and outer barrels. Since the motor is placed above the core barrel, the steel ball cannot be dropped from surface when coring with a PDM. Instead the ball is contained within a special sub, and is released by increasing the flow rate. Coring can begin when the ball closes off the flow through the inner barrel (Fig. 5.8).

SURFACE CHECKS

Before running the motor into the hole, certain checks should be made on surface to reduce the risk of a tool failure downhole. As stated earlier, the amount of bearing wear can be checked by measuring the axial clearance between the bearing housing and the rotating sub. The motor should be replaced if the bearing wear is greater than the maximum limit specified by the manufacturer.

The operation of the dump valve can be checked by physically pushing down on the sleeve to cover up the side ports. When released, the spring should push the sleeve back up to open the ports again. To check the opening and closing of the dump valve with the pumps on, the motor must be connected to the kelly. The motor is lowered through the rotary table until the dump valve is below the rig floor. The motor is then held in position by the rig tongs. By pumping through the motor at a slow rate, it should be possible to see mud exiting through the open ports of the dump valve. By increasing the pump rate, the force on the piston will reach the required point at which it pushes the sleeve down to close off the ports. At this stage the mud is flowing through the tool and the motor should be turning. It is possible to count the number of revolutions to estimate the rpm. Some mud should be seen escaping from the bearing section. The dump valve should open again as the flow rate is reduced.

DRILLING WITH A PDM

After the motor has been successfully tested on surface, the bit and the remainder of the bottom hole assembly can be made up and run in the hole. If a bent sub is being used just above the motor, the drill string must be lowered carefully through any restrictions (e.g. liner laps). Once on bottom the motor should be picked up 1–2 ft before starting the pumps. The hole should be circulated clean before drilling. The off-bottom circulating pressure at the required flow rate should be noted. The motor is then lowered and drilling can commence. The difference between the on-bottom and off-bottom circulating pressure depends on the design of the motor, but is generally 200–500 psi. The WOB and pump pressure should not exceed the manufacturer's recommendations. As the WOB increases, the torque and pump pressure will also increase.

When too much WOB is applied the surface pressure increases sharply, indicating that the motor has stalled and has stopped drilling. The motor must be picked up off bottom and the surface pressure should return to the original circulating pressure. The driller can then go back down to bottom and apply WOB to continue drilling. The driller should try to maintain constant surface pressure by adjusting the WOB. When the bit drills off the pump pressure will decrease and more WOB should be applied. The surface pressure gauge is very important to the driller when using a downhole PDM. The pressure is directly related to the torque and can be used as a WOB indicator. It is therefore fairly easy for the driller to monitor the performance of a PDM since bit speed (rpm) is proportional to pump rate and torque is related to the pressure drop through the motor.

WARNING SIGNS OF PROBLEMS

With any downhole tool there is always the chance of some problem occurring that will affect performance. In the case of a positive displacement motor, a fault may develop within the tool itself or it may cause damage to some other component of the bottom hole assembly (e.g. the bit). In most cases some warning signs can be detected and prompt action should be taken to prevent further damage.

(a) A sudden rise in surface pressure usually indicates that the motor has stalled. If the motor is kept in this stalled condition, serious damage will occur in the stator, preventing a proper seal between the stator and the rotor. If this happens the motor will have to be tripped out of the hole and replaced.

(b) Increased surface pressure may also be caused by debris in the mud clogging the motor and jamming the bearings. Lost circulating material (LCM) can be used, but the size of particles must be carefully chosen. If the motor does become plugged it must be tripped out. (To prevent fill entering the drill string while tripping in, a float sub can be placed above the motor).

(c) Bit wear may be indicated by repeated stalling of the motor owing to a locked cone. A plugged nozzle will cause a more gradual increase in surface pressure.

(d) As the stator and bearings become worn, the pressure drop through the motor decreases. This can be detected by observing a reduction in the surface pressure at which the motor stalls.

(e) If the off-bottom circulating pressure is lower than expected, this may indicate that the dump valve is still open, or that there is a washout in the drill string.

AIR DRILLING

The incentive to use air as the drilling fluid is based on its low density, which results in faster penetration rates. It will also be less damaging to the formation. However, the reduced hydrostatic head also means that formation pressures cannot be controlled and the sides of the borehole cannot be supported. Air drilling, therefore, is limited to competent formations that are essentially dry.

Air compressors are required at surface to circulate the air down the drill string and up the annulus. A rotating air-tight seal is required around the kelly so that the return flow is directed through the "blooey" line, which carries the cuttings and dust away from the rig. If water or oil is found in the formation, the cuttings will become damp and may form mud rings in the annulus. To prevent this, water is mixed with a foaming agent and injected into the air being circulated down the drill string. In this case, a fine mist and foam are circulated out the blooey line.

A positive displacement motor can operate using air or air-mist as the drilling fluid. However, there are certain differences from the normal procedures described earlier:

(i) The side ports of the dump valve should be sealed off, since the air will not exert enough pressure to force down the closing sleeve during drilling.

(ii) One should not circulate through the motor off-bottom. To start drilling tag bottom, apply WOB and then start circulating. This will avoid the problem of excessive torque on the motor.

(iii) Less WOB should be applied in order to prevent stalling.

(iv) A lubricant (liquid soap, gel) should be added to the air stream to prevent damage to the rubber components in the motor.

QUESTIONS

5.1. In a single-lobe motor (1/2 configuration) the rotor diameter is 2.5 in. the eccentricity is 1.125 in. and the rotor pitch is 24 in. At a flow rate of 550 gpm the total pressure drop through the motor is 480 psi. If the motor efficiency is 80%, calculate:

      (a) the rotational speed;

      (b) the torque developed by the motor;

      (c) the power output.

5.2. From the manufacturer's specifications a 3-stage, single-lobe motor has a rotor diameter of 4.555 in, eccentricity 1.5 in. and a rotor pitch of 62 in. The available torque is 5666 ft lb. Calculate the pressure drop per stage in the motor.

5.3. A directional well is being kicked off using a PDM and a bent sub. The driller is maintaining a fairly constant surface pressure of 2700 psi and a WOB of 20,000 lb. As more WOB is applied there is a sudden rise in the surface pressure and the ROP falls to zero. Explain why this happens and what course of action the driller should take.

5.4. For a 9½ in. PDM (single-lobe) whose rotor diameter is 2.5 in., length of stage 50 in. and eccentricity 1.125 in., plot power versus pressure drop (200–400 psi) for a range of flow rates (200–400 gpm).

5.5. In planning a kick-off the operator has selected the following parameters:

      flow rate = 550 gpm

      bit speed = 150 rpm

      bit pressure drop = 900 psi

      drill string + annulus pressure drop = 1800 psi

      maximum surface pressure = 3000 psi

      The downhole motors available are as shown in Table 5.1. Select the motor that fits these requirements, and calculate the power output.

5.6. What are the advantages of using a motor with a bent housing rather than a bent sub, and for what applications might a bent housing be used?

FURTHER READING

"New low speed, high torque motor experience in Europe", A. J. Beswick and J. Forrest, S.P.E. paper no. 11168.

"New downhole motor develops high torque for increased penetration rates", C. McCabe, *Ocean Industry*, June 1982.

"Performance drilling optimisation", H. Karlsson, T. Brassfield and V. Krueger, S.P.E./I.A.D.C. paper no. 13474.

Hydraulic Downhole Drilling Motors, W. Tiraspolsky, Gulf Publishing Company, 1985.

"PDM vs turbodrill—a drilling comparison", F. V. deLucia and R. P. Herbert S.P.E. paper no. 13026.

Chapter 6

TURBODRILLS

Even before directional drilling became established, some early development work had been carried out on downhole turbines that could be used to drive the bit. These early tools were aimed at providing an alternative to conventional rotary methods in which the entire drill string had to be rotated from surface. The turbodrill did offer several advantages over rotary drilling for straight holes.

(a) Rotary torque was developed at the bit where it was actually required;
(b) The turbodrill could develop more power than the rotary system could transmit to the bit.
(c) The turbine could produce much faster rotational speed.
(d) Since the drill string did not have to rotate, there was less wear on the drill pipe and casing.

In 1873, C. G. Cross of Chicago patented a single-stage turbine, but there is no record of this tool actually being used. It was not until the 1920s that interest in turbodrills was revived with development work being carried out mainly in the USA and USSR. In 1924, a single-stage turbine was developed by Kapelyushnikov and tested at Baku on the Caspian Sea. Multistage turbines were first used by Scharpenberg in California in 1926. One such model was 9 in. in diameter and had 30 stages. It was capable of producing 92 hp at 700 rpm with a flow rate of 550 gpm. However, turbodrilling made little impact in the USA and field trials were abandoned around 1950. Conventional rotary methods dominated drilling operations in the USA, although Dresser did import some Russian turbodrills in the mid-1950s. In the USSR, however, further developments took place, until by 1954 over 80% of Russian oil wells were being drilled by turbines. There was also some interest in Western Europe, especially in France where Neyrfor began making downhole turbines in 1956.

One of the major problems in using turbodrills was in controlling the rotational speed, which was too fast for conventional rock bits. Early Russian turbodrills had reducers that geared down the speed from 2000 rpm to about 30 rpm. These reducers, however, quickly wore out and proved unreliable. In the USA, it was clear that improvements were required both in the design of turbines and in the design of drilling bits before turbodrilling would be successful. Although turbodrilling could not compete in economic terms with rotary drilling in straight holes, there were some advantages to be gained in directional holes.

With the great expansion of directional drilling activity during the 1960s and 1970s, Western countries began taking turbodrilling more seriously. As exploration and production moved into offshore areas such as the Gulf of Mexico and the North Sea, drilling costs escalated. The high cost of drilling directional wells from fixed platforms provided the incentive to improve drilling efficiency. Turbodrills demonstrated that penetration rates could be substantially increased. With further improvements in turbine design and the introduction of polycrystalline diamond (PDC) bits, turbodrilling became more popular. The more powerful mud pumps on offshore rigs delivering higher discharge pressures also contributed to the success of turbodrilling.

Turbodrills may be used for both straight and directional wells in competition with conventional rotary methods. Many operating companies will analyse the costs involved and compare rotary and turbine performance before deciding which method should be used over certain intervals. The turbine may consist of several sections and be up to 50 ft in length (Fig. 6.1). Shorter turbodrills may be used for kick-offs with a bent sub, but positive displacement motors are usually preferred for these operations. The major application of turbodrills is to drill the long tangential section of a deviated well, since this is where substantial savings can be made.

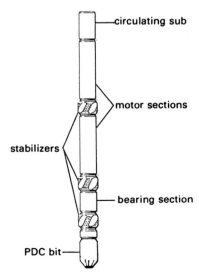

Fig. 6.1. Multisection turbodrill (courtesy of Eastman-Christensen).

MOTOR DESCRIPTION

The turbine motor consists of a series of rotors and stators. The rotors are blades that are mounted on a vertical shaft, while the stators are fixed to the body of the turbodrill (Fig. 6.2). Each rotor–stator pair is called a "stage". The number of stages may vary between 25 and 250, depending on the application. The pressure drop through each stage should be constant. The number of stages is therefore limited by the total head available. Each stage also contributes an equal share of the total torque and the total power developed by the turbine.

As the drilling fluid is pumped through the turbine, the stators deflect the flow of mud against the rotors, forcing them to turn the vertical shaft in a clockwise direction. Thrust bearings are used to withstand the axial loads, while radial bearings centre the shaft while it is rotating. After passing through the stages of the turbine, a small percentage of the flow is diverted through the lower bearing for lubrication. The rest of the flow is channelled into the hollow lower section of the shaft and down through the bit nozzles (Fig. 6.3).

The upper end of the shaft is free to rotate, while the lower end is connected to the bit. At the top of the turbodrill there is a box connection made up on the end of the drill string. The turbine blades are usually made of carbon steel that contains a small percentage of alloying metal (chromium). The blades may be positioned within an inner and outer ring to form a wheel. The profile of the blades has an important effect on the pressure drop through the tool and on the turbine's overall performance. All turbine components must be robust to withstand the expected down-hole conditions. Modern turbodrills should be capable of 200 hours of continuous drilling.

The minimum clearance between stator and rotor is of the order of millimetres. Any solid particles or debris in the mud could easily cause clogging of the motor. A sand content of 2% or more could also cause

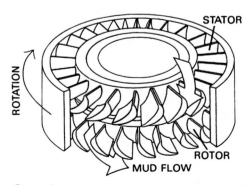

Fig. 6.2. Operation of stators and rotors (courtesy of Drilex).

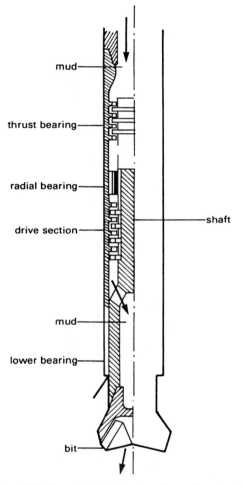

Fig. 6.3. Longitudinal section through turbodrill.

excessive wear of the turbine blades. Lost circulation material cannot be allowed through the turbine and must be diverted through a circulating sub installed above the turbine for this purpose. A straining device must be fitted to protect the turbine from any other debris in the mud. This device (mud screen) can be placed just above the turbine, but this makes it difficult to retrieve if it needs to be emptied. It may also obstruct the passage of wireline tools. It is generally more convenient to install the mud screen in a more accessible place at the kelly or in the top section of drill pipe.

BEARINGS

In turbodrilling operations the life of the bearings is a critical factor. One of the common problems is for the bearings to fail, forcing the operator to pull the tool out of the hole before the bit is worn. Bearing performance has been improved by using different bearing designs and installing them at different positions within the tool. The types of bearing normally used are described as follows.

Thrust Bearings

These resist the axial load exerted on the turbine. The simplest type of thrust bearing consists of metal discs that slide on a bearing surface of elastomer or synthetic rubber. The metal discs are attached to the shaft while the elastomer is fixed to bearing supports on the inside of the body of the tool. Since axial loads may be applied in either direction, the elastomer material must be used on both the upper and lower sides of the bearing support (Fig. 6.4). There are also channels through which the drilling fluid lubricates the bearing. The resisting torque developed in the thrust bearing depends on several factors, including the surface area, the axial loading and the coefficient of friction. Ball bearings and rollers can also be used in thrust bearings. The resisting torque is much less in this type of bearing.

Thrust bearings can be located between the motor section and the bit, thus isolating the turbine blades from vibration and shock loads. In multisection turbines, thrust bearings may be installed within each section.

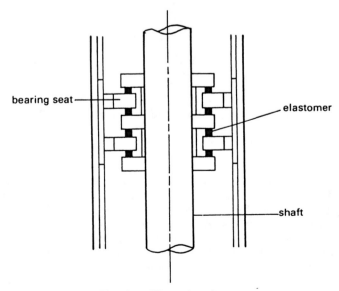

Fig. 6.4. Thrust bearings.

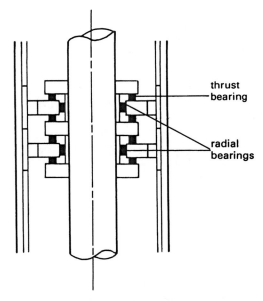

Fig. 6.5. Combined radial and thrust bearings.

Radial Bearings

The purpose of radial bearings is to centralize the drive shaft. Elastomer pads on the inside of the bearing supports and a metal sleeve on the shaft are fitted such that a small clearance (0.5 mm) exists between them. This type of bearing is also lubricated by the drilling mud. It is possible to combine radial and thrust bearings as shown in Fig. 6.5. Radial bearings may be located along the length of the shaft at intervals of 6–10 ft.

Lower Bearing

The function of the lower bearing is to centralize the lower part of the drive shaft and to resist the bending stresses exerted on the turbine while drilling. In order to fulfil these requirements, both ball bearings and roller bearings may be used. The bearing is lubricated by diverting 5–10% of the total mud flow through the bearing and out into the annulus. Longer bearing life can be achieved by designing sealed bearings that are lubricated by oil.

PERFORMANCE CHARACTERISTICS

The turbodrill is essentially an axial flow machine in which fluid energy is transferred to the rotor shaft. There is no appreciable radial flow, since the radius remains constant throughout. The absolute velocity of the fluid as it passes through the turbine can be divided into two components (Fig. 6.6).

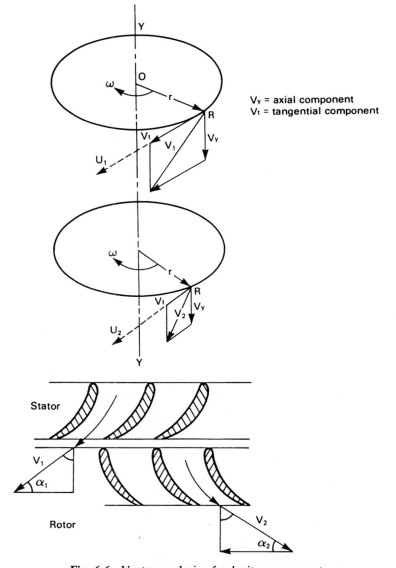

Fig. 6.6. Vector analysis of velocity components.

The component parallel to the axis YY is denoted by V_y, while the tangential component (i.e. at right angles to the radius OR) is denoted by V_t. Only the V_t component does useful work in turning the rotor. The mud is deflected by the stator as it enters the turbine. Although there is a change of direction, no work is done by the fluid on the stator, since the stator remains fixed.

Let V_1 denote the initial velocity as the fluid strikes the rotor. The angle at which V_1 acts with respect to the direction of rotor movement is α_1. Similarly, let the final velocity of the fluid be V_2 acting at the angle α_2. From the vector diagrams in Fig. 6.6, it can be seen that $V_{t1} = V_{y1} \cot \alpha_1$ and $V_{t2} = V_{y2} \cot \alpha_2$. Since $V_{y1} = V_{y2} = V_y$ for continuity of axial flow through the turbine, the change in velocity can be written as

$$V_{t2} - V_{t1} = V_y \left(\cot \alpha_2 - \cot \alpha_1 \right)$$

The rate of change of momentum can be found by multiplying the velocity change by the mass flow rate (ρQ) i.e. the fluid density multiplied by the flow rate. From Newton's second law, therefore, the force F applied to the rotor is found by:

$$F = \rho Q (V_{t2} - V_{t1})$$

The torque T produced by this force is given by:

$$T = Fr = \rho Q r (V_{t2} - V_{t1})$$

Power P can be determined from $P = T\omega$, where ω is the angular velocity in radians per second. Since the tangential velocity U_t of the rotor blade is given by $U_t = \omega r$, the power equation can be written as

$$P = \rho Q (U_{t2} V_{t2} - U_{t1} V_{t1})$$

To obtain the pressure drop Δp the power is divided by the flow rate, giving

$$\Delta p = \rho (U_{t2} V_{t2} - U_{t1} V_{t1})$$

Since r is constant, $U_t = U_{t2} = U_{t1}$ and the equations for torque, power and pressure drop become

$$T = \rho Q r (V_{t2} - V_{t1})$$
$$P = \rho Q U_t (V_{t2} - V_{t1})$$
$$\Delta p = U_t \rho (V_{t2} - V_{t1})$$

Notice that T, P and Δp are all directly proportional to the mud density ρ. From the above equations it can be shown that T, P and Δp are proportional to some power of Q, i.e. $T \propto Q^2$, $P \propto Q^3$ and $\Delta p \propto Q^2$. These relationships can be used to calculate the changes in T, P and Δp when the flow rate changes from Q_1 to Q_2. If T_1, P_1 and Δp_1 are the values corresponding to Q_1, the new values corresponding to Q_2 can be calculated from

$$T_2 = T_1 \left(\frac{Q_2}{Q_1} \right)^2$$

$$P_2 = P_1 \left(\frac{Q_2}{Q_1} \right)^3$$

$$\Delta p_2 = \Delta P_1 \left(\frac{Q_2}{Q_1} \right)^2$$

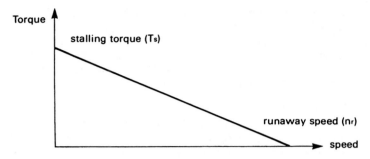

Fig. 6.7. Torque versus rotational speed.

It is important in assessing the performance of a turbodrill to consider how certain key parameters are changing with respect to rotational speed. The maximum rotational speed occurs when there is no resistance to the flow of fluid through the turbine. This is known as the "runaway speed" and corresponds to zero resisting torque. At the other extreme, the turbine stalls when the rotors are prevented from turning. The resisting torque therefore reaches a maximum at the stall condition. The relationship between torque and speed is approximately linear, as shown in Fig. 6.7. The equation of this straight line can be expressed as

$$T = mn + c$$

where the gradient

$$m = -\left(\frac{\text{stalling torque, } T_s}{\text{runaway speed, } n_r}\right)$$

and the intercept $c = T_s$. The torque can therefore be given as

$$T = -\left(\frac{T_s}{n_r}\right)n + T_s \qquad \text{or} \qquad T = T_s\left(1 - \frac{n}{n_r}\right)$$

The power developed by the turbine can be determined from $P = T\omega$. The angular velocity ω can be related to the rotational speed, since $\omega = (2\pi/60)n$, where n is measured in revolutions per minute and ω is measured in radians per second. Substituting for T gives power in terms of rotational speed:

$$P = \frac{\pi n}{30}T_s\left(1 - \frac{n}{n_r}\right)$$

To convert power to oilfield units (hp), multiply this by 1/550. Power therefore is a quadratic function of rotational speed as can be seen from Fig. 6.8. To find the optimum speed for which power reaches a maximum, the derivative dP/dn is set equal to zero.

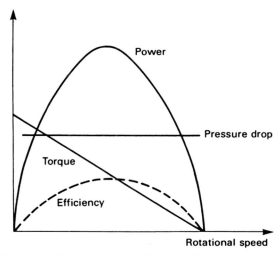

Fig. 6.8. Typical performance curves for a turbine for constant flow rate.

From the above equation

$$P = \left(\frac{\pi T_s}{30}\right)n - \left(\frac{\pi T_s}{30n_r}\right)n^2$$

$$\Leftrightarrow \frac{dP}{dn} = \frac{\pi T_s}{30} - \left(\frac{2\pi T_s}{30n_r}\right)n = 0$$

Therefore the optimum rotational speed $n_0 = n_r/2$. The maximum power is hence obtained when the rotational speed is exactly half of the runaway speed. Notice that at the optimum speed the torque will be exactly half the stalling torque, i.e. $T = T_s/2$.

In conventional turbines the pressure drop through the tool varies only slightly with rotational speed (Fig. 6.8). For operational purposes, pressure drop can be assumed to remain steady for a constant flow rate.

The overall efficiency E of a turbodrill is the product of three components:

$$E = E_h \times E_v \times E_m$$

The hydraulic efficiency term E_h takes into account pressure variations and entry and exit losses. The volumetric efficiency E_v accounts for the fluid that passes through the clearances between rotors and stators and therefore does not contribute to the power output. The mechanical efficiency E_m accounts for losses due to friction in the bearings. Theoretically the maximum efficiency should coincide with the optimum rotational speed (Fig. 6.8). In practice this may not be the case, since the torque–speed relationship is not perfectly linear. The effective power output in terms of flow rate and pressure drop can be expressed as:

$$P = \frac{EQ\,\Delta p}{1714}$$

where P = power (hp)
       Q = flow rate (gpm)
       Δp = pressure drop through turbine (psi)
       E = overall efficiency.

SURFACE CHECKS ON THE TURBODRILL

The service company supplying the turbodrill will have tested the tool in their workshop before transporting it to the rig. There should also be an experienced service company engineer on the rig to check the tool before it is run into the hole and to advise on how best to operate the tool downhole. It is good policy to carry out a few surface checks on the tool before running it, rather than discover any faults when the tool is on the bottom.

(a) When making up the tool on the rig, careful attention should be paid to the recommended make up torque. Multisectioned tools should be checked for vertical alignment.

(b) The axial clearance between rotor and stator should be measured at the lower end of the turbine. The distance between a reference mark on the shaft and the end of the lower bearing must be measured with a caliper. This measurement is carried out with the turbine freely suspended, and again when resting on the rig floor (Fig. 6.9). The actual measurements should be compared with the manufacturers' acceptable wear limits. Also, the radial clearance between the rotor and the casing of the turbine should be checked. Both axial and radial clearances should be very small (of the order of millimetres).

(c) A rotation test should be carried out to ensure that the turbine's operation is satisfactory. The usual procedure is to make up the kelly and pump through the tool while the tool is held in the tongs. The bottom of the tool should be lowered through the rotary table to prevent mud spilling over the rig floor. If a tachometer is available, a measurement of rotational speed for various different flow rates can be made. Checks should also be made for undue noise, vibration or leakages during the rotation test.

After the tool has been successfully tested on surface it should run in the hole at a steady rate. Any sudden jerks or collisions with the casing or open hole may cause damage to the turbine and the bit. Once it is on bottom, weight should be applied gradually as the bit begins to drill.

DRILLING WITH A TURBODRILL

There is still a considerable "art" involved in drilling with a turbine and it is usually the directional driller's responsibility to decide how the tool should be operated. The service company that supplied the tool should provide

(a)

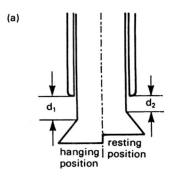

(b)

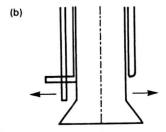

Fig. 6.9. Checking turbine clearances. Axial clearance is measured at the lower end of the turbodrill with a sliding caliper between the end of the lower bearing and the shaft shoulder. (a) The measurements are taken in both the hanging and the resting position. The clearance is $d_1 - d_2$. For a new turbodrill the axial clearance is about 1 mm. The maximum permissible wear given by the manufacturer is around 5–6 mm. (b) Radial clearance is measured with a rule and a graduated gauge. Radial clearance must be 1 or 2 mm maximum (check with manufacturer for wear limits).

sufficient technical information concerning the expected performance of their particular turbine (e.g. performance charts or recommendations for flow rates, pressure drops and torque). A typical performance chart is shown in Fig. 6.10. It is also the directional driller's job to see that the well is drilled according to the planned trajectory. He must therefore be familiar with the directional behaviour of the turbine and make up the bottom hole assembly accordingly. In particular he must take account of the turbine's reactive torque which will cause left hand walk at the bit.

If turbodrilling is to compete with conventional rotary methods, the extra cost of using the turbine must be justified by ensuring greater penetration rates than would otherwise be possible. To optimize the operation, the following factors must be considered.

Flow Rate

The flow rate is proportional to the rotational speed of the turbine when it is checked on surface with a tachometer. When the turbine is downhole,

Dimensions:

Recommended Hole Size	Nominal Tool Size	Actual Tool O.D.	Thread Connections Top Box	Bit Sub	BS	MS	TMS	CS	Length	Weight
inch	inch	inch			ft	ft	ft	ft	ft	lbs
12¼ – 17½	9½	9⅝	7⅝" API Reg.	6⅝" API Reg.	16.1	17.1	16.4	2.6	69.3	13900

Performance data:

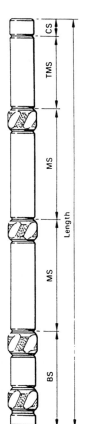

Pump Rate	Bit Speed	Power Output	Torque	Pressure Drop	Thrust Load
GPM (US)	RPM	HP	ft lbf	PSI	1000 lbf
475	527	226	2249	1162	35
500	554	263	2492	1288	39
525	582	305	2748	1420	43
550	610	350	3016	1558	47
575	637	400	3296	1703	51
600	665	455	3589	1855	56
625	693	514	3894	2012	60
650	**721**	**578**	**4212**	**2177**	**65**
675	748	648	4542	2347	70
700	776	722	4885	2524	76
725	804	803	5240	2708	81
750	831	889	5608	2898	87

Mud weight
10 lbs/gal

Pump Rate	Bit Speed	Power Output	Torque	Pressure Drop	Thrust Load
GPM (US)	RPM	HP	ft lbf	PSI	1000 lbf
475	527	271	2699	1395	42
500	554	316	2991	1546	46
525	582	366	3297	1704	51
550	610	420	3619	1870	56
575	637	480	3955	2044	61
600	665	546	4307	2226	67
625	693	617	4673	2415	72
650	721	694	5055	2612	78
675	748	777	5451	2817	84
700	776	867	5862	3029	91
725	804	963	6288	3250	97

Mud weight
12 lbs/gal

Pump Rate	Bit Speed	Power Output	Torque	Pressure Drop	Thrust Load
GPM (US)	RPM	HP	ft lbf	PSI	1000 lbf
475	527	316	3149	1627	49
500	554	369	3489	1803	54
525	582	427	3847	1988	60
550	610	491	4222	2182	65
575	637	561	4615	2385	71
600	665	637	5025	2597	78
625	693	720	5452	2817	84
650	721	810	5897	3047	91
675	748	907	6359	3286	98

Mud weight
14 lbs/gal

CS – Circulating Sub
TMS – Top Motor Section
MS – Motor Section
BS – Bearing Section

Fig. 6.10. Example of performance charts. (On diagram: CS = circulating sub; TMS = top motor section; MS = motor section; BS = bearing section.) (Courtesy of Eastman-Christensen.)

however, the WOB will affect the rotational speed. An MWD tool with an rpm sensor must be used for an accurate measurement of rotational speed under downhole conditions.

Since the power developed by the turbine is proportional to the cube of the flow rate, a very small increase in flow rate will produce a large increase in power. However, there are other limitations on flow rate that must be considered.

(a) For lifting cuttings, the annular velocity must not be allowed to fall below a certain minimum.

(b) In areas of weak or unconsolidated formations, the annular velocity must be kept below a certain maximum to prevent erosion and borehole instability.

(c) The operating company may wish to use the flow rate recommended by the bit manufacturer.

Mud Weight

The mud density is directly related to the pressure drop, power and torque developed by the turbine. However the mud weight will usually be determined from the expected formation pressures. The hydrostatic pressure of the mud column must equal or be slightly greater than the formation pressure. If this overbalance is too great, the penetration rate will be reduced owing to chip hold-down effect.

Other mud properties, such as sand content, will also affect turbine performance. A sand content of 2% or more will cause abrasion damage to the turbine blades. The friction coefficient in the thrust bearings will also be affected, causing bearing wear.

Pressure Drop through Turbine

To maximize the hydraulic horsepower, the optimum pressure drop across the bit should be about 67% of the total pressure available at the pump. However, the pressure drops through the various other drill string components, including the turbine, may be greater than the 33% required for optimization. A large pressure drop through the turbine may reduce the flow rate, such that the annular velocity is not sufficient to clean the hole.

WOB–RPM

The WOB-rpm relationship is a vital factor in determining penetration rate and bit wear. In general, the higher the WOB the greater the rate of penetration (ROP), provided that hole cleaning is adequate. In turbodrilling, however, too much WOB will cause stalling. For turbodrilling operations, the two most common types of bits used are natural diamond bits or PDC bits. It has been shown that natural diamond bits perform better at higher WOB and lower rpm compared with PDC bits which perform better with lower WOB and higher rpm.

Drill String Rotation

Drill string rotation during performance drilling (e.g. when drilling the tangential section of a directional well with a turbine) offers several advantages.

(a) It reduces the risk of differential sticking.
(b) It allows a smoother application of weight to the bit.
(c) It improves mud circulation in the annulus, which helps to prevent cuttings settling.

In North Sea operations, the drill string is rotated at around 100 rpm while the turbine speed is around 700 rpm or more. Under these conditions the extra power provided by the rotary table does not have an appreciable effect on turbodrill performance. Owing to the interaction of drill string rotation and turbine rotation, the power delivered by each is not directly cummulative. Drill string rotation tends to reduce the left-hand walk at the bit caused by reactive torque.

OPTIMIZATION WHILE DRILLING

Under normal circumstances with conventional turbines, the only parameters which can be controlled by the driller are flow rate and WOB. The optimum WOB can be determined from a drill-off test. But this can be time-consuming and applies only to one particular formation. Field tests have shown that ROP is proportional to the power developed at the bit (Fig. 6.11). The most accurate means of determining the downhole power at the bit is to use the following formula:

$$P = \frac{T \times n}{5252}$$

where T = torque (ft lb)
      n = rotational speed (rpm)
      P = power (hp).

However, T and n would have to be measured under downhole conditions, which is only possible with the use of MWD tools (unlike the case of a PDM, surface pressure does not give an accurate indication of T and n). If downhole sensors are not available, power can be calculated from

$$P_2 = P_1 \frac{MW_2}{MW_1}\left(\frac{Q_2}{Q_1}\right)^3$$

where P_1 = power developed under test conditions with mud weight MW_1 and flow rate Q_1. This information is supplied by manufacturers. P_2 is the power developed under field conditions using mudweight MW_2 and flow rate Q_2.

It can be seen from the characteristic curves shown in Fig. 6.8 that

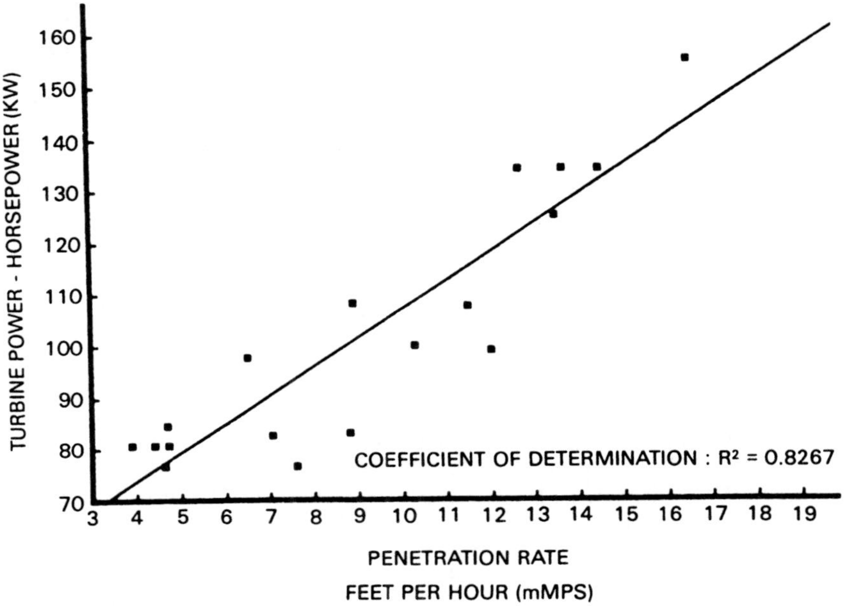

Fig. 6.11. Turbine power output versus rate of penetration (ROP) (after Mason).

maximum power occurs at an optimum rotational speed equal to half the runaway speed of the turbine. For maximum ROP, therefore, the turbine must be operated at this speed. The runaway speed, published in manufacturers catalogues, may be inaccurate since the test might have been carried out using water instead of mud. It is better to check the runaway speed by rotating the turbine off bottom at a given flow rate and to rely on the MWD tool with an rpm sensor to determine the runaway speed. The turbine is then lowered to bottom and WOB gradually applied. The introduction of downhole tachometers sending data back to surface in real time has opened up a great opportunity for optimizing turbodrill performance. Knowing the actual runaway speed, the driller can adjust the WOB until the optimum rotational speed is reached. By monitoring the downhole rpm continuously and adjusting WOB accordingly, optimum conditions can be maintained for maximum penetration rate.

While drilling with a turbine, the driller must be alert for signs that indicate a downhole problem with the tool. The surface pressure gauge will register an increase if there is any blockage in the turbine or in the bit. Likewise, a reduction in surface pressure may indicate a leakage somewhere in the tool. The introduction of downhole rpm and torque sensors gives the driller an even better appreciation of downhole problems.

The effectiveness of turbodrilling operations is further enhanced by improving the hydraulic capacity of some other components in the drilling system.

(a) Sufficient pump capacity: The mud pumps on the rig must be able to deliver high flow rates at high discharge pressure (over 4000 psi in some cases).
(b) Selection of drill pipe sizes: Small-ID drill pipe will restrict the flow rate and increase pressure drops through the drill string. Larger-diameter drill pipe will allow better turbine performance.
(c) Selection of bottom hole assembly: When turbodrilling with a PDC bit, WOB can be reduced. The number of drill collars should therefore also be reduced, or should have larger ID.

TURBODRILLING IN THE NORTH SEA

The high operating costs involved in offshore drilling have encouraged the use of turbodrills to reduce drilling time. Downhole turbines were introduced in the North Sea in 1968 and have been used both in straight and deviated wells. The combination of turbines and polycrystalline diamond compact (PDC) bits has produced some very successful results in the long tangential sections of directional wells. PDC bits have self-sharpening cutters made from artificial diamond bonded to tungsten carbide (Fig. 6.12). These cutters produce a highly efficient shearing action. They are most effective in drilling through soft–medium formations such as the claystone found in the cretaceous section of some of the fields in the northern North Sea. Owing to the sensitivity of these formations to water, an oil-based mud is generally used for drilling the tangential section.

A typical bottom hole assembly for such an application is shown in Fig. 6.13. The position of the stabilizers can be altered to make the assembly hold, drop or build angle as required. The near-bit stabilizer helps to prevent the bit from wandering off course. Since less weight on bit is required when drilling with a PDC bit, fewer drill collars are required and so the weight of the drill string is less than that for a rotary assembly.

Typical operating parameters for this kind of assembly are given below:

Hole size	$12\frac{1}{4}$ in. ($9\frac{1}{2}$-in. turbine)
WOB	4000–8000 lb
RPM at bit	900 rpm (120 at rotary)
Flow rate	600 gpm
Surface pressure	4250 psi
ROP	60–80 ft per hour

To compare the performance of turbines with conventional rotary methods the economics must be considered. The following equations can be used to calculate cost per foot drilled.

For rotary drilling

$$\text{Cost/ft} = \frac{C_b + C_r(T_d + T_t)}{F}$$

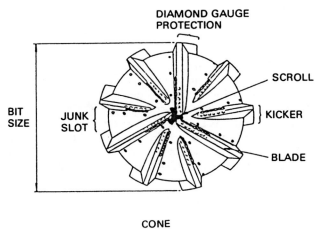

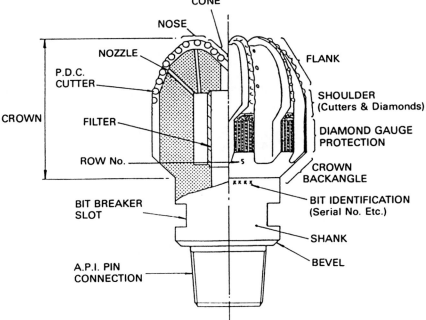

Fig. 6.12. Elements of a polychrystalline diamond bit (courtesy of Drilling and Service).

where C_b = bit cost ($)
     C_r = rig operating cost ($/h)
     T_d = drilling time (h)
     T_t = tripping time (h)
     F = interval drilled (ft).

For turbodrilling

$$\text{Cost/ft} = \frac{C_b + C_r(T_d + T_t) + C_t T_d}{F}$$

where C_t = cost of using the turbine. (This is taken to include the hire of the service company engineer as well as the turbine itself.) In some cases the cost of running the turbine also includes the trip time, i.e. $C_t(T_d + T_t)$.

Consider the following example based on typical North Sea figures. With conventional rotary methods a rock bit costing $3500 drilled 350 ft in 15 hours. In the same formation a turbine with a PDC bit costing $40,000 drilled 2000 ft in 40 hours. The additional cost of using the turbine was $200 per hour. Assume rig cost ($1500 per hour) and trip time (5 hours) to be the same for both runs.

$$\text{Rotary cost/ft} = \frac{3500 + 1500(15 + 5)}{350} = \$95.7$$

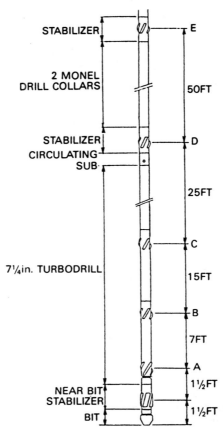

Fig. 6.13. Example of a turbodrill assembly as used in the North Sea.

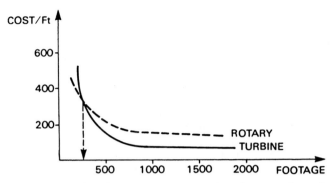

Fig. 6.14. Example of a break-even curve comparing the cost of running a turbodrill as opposed to rotary methods. If the footage is less than 250 ft conventional rotary drilling is more economical.

$$\text{Turbine cost/ft} = \frac{40,000 + 1500(40 + 5) + (200 \times 40)}{2000} = \$57.8$$

Under these conditions, therefore, turbodrilling is more economic than rotary methods. It is not only the increased ROP that makes turbodrilling attractive. Longer bit runs with PDC bits mean that fewer bits will be required and therefore fewer trips. Over long sections of homogeneous soft–medium formations, this results in substantial lowering of overall drilling costs despite the additional expense of hiring turbines and using PDC bits. In some North Sea fields a reduction of more than 50% in drilling costs has been achieved. Break-even curves can be drawn up to specify the conditions under which turbodrilling would be economically justified (Fig. 6.14). In addition to the direct savings in the drilling operation, there are also indirect savings in terms of reduced wear on the drill string and surface equipment.

QUESTIONS

6.1. At a flow rate of 600 gpm a turbodrill produces a torque of 3589 ft lb at a speed of 665 rpm. Calculate:
(a) the power output;
(b) the total pressure drop through the turbine at this flow rate assuming an efficiency of 70%.
6.2. The following information on turbine performance has been taken from the manufacturer's tables for a mud weight of 10 ppg.

Flow rate (gpm)	Speed (rpm)	Torque (ft lb)	Pressure drop (psi)	Power output (hp)
500	554	2492	1288	263

If the mud weight is increased to 11.5 ppg and the flow rate becomes 620 gpm calculate the new values of speed, torque, pressure drop and power.

6.3. Describe the surface checks that should be made on a turbodrill before it is run in the hole.

6.4. During laboratory tests on a particular turbine the stalling torque was found to be 1800 ft lb and the runaway speed was 1500 rpm.
    (a) Plot the theoretical torque versus speed graph.
    (b) Plot the corresponding power versus speed graph.
    (c) Determine the optimum speed for maximum power.

6.5. To compare the cost of turbodrilling with conventional rotary drilling over a certain interval, construct a breakeven curve using the following data.
    (i) For rotary, average ROP = 5 ft/h, C_b = $5000 (mill tooth).
    (ii) For turbine, average ROP = 20 ft/h, C_b = $18,000 (PDC bit), cost of turbine = $400 per hour (rotating on bottom).
    Assume a rig cost of $500 per hour and a trip time of 16 hours in each case. Calculate the cost/ft for footages ranging between 50 ft and 300 ft. What is required footage to make the turbodrill economically viable?

6.6. Compare the major advantages and disadvantages of turbodrills versus positive displacement motors for directional drilling applications.

FURTHER READING

"Turbodrill, PDC bits spell success in Mobil's Statfjord", R. W. Turnbull, *P.E.I.*, October 1982.

"The versatility of the turbodrill in North Sea drilling", R. Powell, G. Cooke and A. Hippman, Europec Conference, London 1980 (EUR 245).

"The use of MWD for turbodrill performance optimisation as a means to improve rate of penetration", H. J. deBruijn, A. J. Kamp and J. C. M. van Dongen, S.P.E. paper no. 13000.

"Optimising diamond bit/turbine drilling performance", C. M. Mason, S.P.E. paper no. 12613.

Chapter 7

DIRECTIONAL SURVEYING

In both straight and deviated holes the position of the wellbore beneath the surface must be determined as the well is being drilled. This requires the use of surveying instruments that are able to measure the hole inclination and direction at various depths along the course of the well. The position of the wellbore relative to the surface location can be calculated from the cumulative survey results.

The early borehole surveying instruments were fairly crude and inaccurate. The need for more reliable surveying tools became apparent with the expansion of directional drilling activities. Modern survey tools make use of sophisticated solid-state devices to provide the high degree of accuracy required.

The objectives in directional surveying are as follows:

(a) to monitor the actual wellpath as drilling continues to ensure that the target will be reached;
(b) to orient deflection tools in the required direction when making corrections to the well path;
(c) to ensure that the well being drilled is in no danger of intersecting an existing well nearby;
(d) to determine the true vertical depths of the various formations that are encountered to allow accurate geological mapping;
(e) to determine the exact bottom hole location of the well for the purposes of monitoring reservoir performance, and also for relief well drilling;
(f) to evaluate the dog-leg severity along the course of the wellbore

The surveying of oil wells began in the 1920s when it was discovered that many so-called vertical wells were in fact deviated by up to 30° from true vertical. These large deviations accounted for many of the dry holes encountered on some of the early oilfields. More effort was put into choosing suitable bottom hole assemblies, and changing drilling parameters to ensure that the hole deviation was reduced to acceptable limits.

As directional drilling became more common, surveying assumed a greater importance than it had for straight holes. It was possible by measuring both inclination and direction at various depths, to plot the course of the well towards the target. By the 1960s surveying tools and practices were well established, but the high costs involved in offshore drilling made operators look closely at the time taken to run surveys. In drilling a directional well from an offshore platform, surveying could account for 10% of the total drilling time. Running single shots on wireline for surveying the hole and orienting deflecting tools was therefore very expensive. This provided the incentive for more sophisticated methods such as steering tools, and more recently MWD, to be introduced. The improvement in surveying techniques has been accompanied by a better understanding of directional control. Continuous monitoring gives the directional driller the ability to assess the effects of changing drilling parameters on borehole inclination and azimuth. With an MWD tool, measurements of azimuth, inclination and toolface can be sent to surface in a matter of minutes.

A wide variety of surveying instruments has evolved over the years. The early instruments relied on a very simple mechanism to measure and record the necessary angles. Although some of the older survey tools have now been abandoned, the single shot instrument is still sometimes used to check the results of more advanced instruments.

ACID BOTTLE

The earliest form of surveying tool used in the oil industry was the "acid bottle". The technique had been used in the mining industry from about 1870. It is based on the simple principle that the free surface of a liquid always remains horizontal regardless of how its container is positioned. In this particular instrument the container is a glass cylinder and the liquid is hydrofluoric acid. If the instrument is allowed to rest in an inclined position for a certain period of time the acid will react with the glass and leave a mark on the side of the cylinder indicating the horizontal surface. The distance between the mark and the acid's original position when the cylinder was level can be used to calculate the inclination angle (Fig. 7.1). The strength of the acid must be chosen carefully to etch a sharp distinct line on the glass within a reasonable length of time.

The instrument was lowered down the drill string on wireline until it rested on top of the bit or on a baffle plate at some point above the bit. The acid bottle was left in this position for about 30 minutes to allow the reaction to take place. The motion of the acid during running in and pulling out prevented any other lines being etched on the glass. The glass was inspected back at the surface and the angle of inclination was determined.

To measure the hole direction, an additional compartment was required containing gelatine and a magnetic compass needle. The compass needle was free-floating and aligned itself with Magnetic North. It was held in this position by the gelatine. The direction of the deviated well could therefore be referenced to Magnetic North.

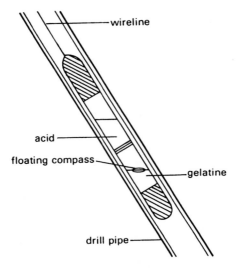

Fig. 7.1. "Acid bottle".

The major disadvantage of the acid bottle technique was that the acid did not always leave a distinct line to show the interface. In reading the mark some allowance had also to be made for capillary effects.

PHOTOMECHANICAL DEVICES

By 1930 the inaccuracy and inconvenience of the acid bottle technique led to many attempts to design a better instrument. The primary requirements were:

(a) to find an accurate and foolproof method of measuring inclination and direction at the bottom of the hole;

(b) to devise a system of recording these angles so that they could be interpreted easily when the instrument was brought back to surface.

The majority of the devices produced used a plumb bob to measure inclination and a compass to measure direction. A camera was incorporated into the instrument to record the angles downhole. The camera photographed the angle-measuring devices when in a stationary position at the bottom of the drill string. When the instrument was recovered at surface, the small photographic disc was developed and survey results were read off directly from the picture. This type of instrument became known as the "single shot". Owing to its reliability it became widely accepted as the standard surveying tool in the oil industry. The instrument is run on an assembly as shown in Fig. 7.2. Its main components are as follows:

(a) a guiding nose to locate the assembly in the baffle plate;

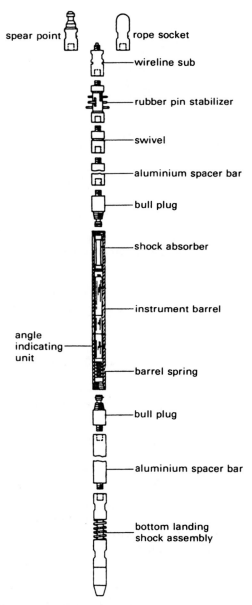

Fig. 7.2. Surveying assembly run on wireline (courtesy of Eastman-Christensen).

(b) shock absorbers above and below the instrument barrel to prevent shock loads while running in the hole;

(c) rubber stabilizers to centralize the instrument inside the drill string;

(d) sinker bars to add enough weight to allow the entire assembly to free-fall;

(e) spacer bars to position the compass at the optimum point within the non-magnetic collar (to reduce the risk of magnetic interference);

(f) a rope socket for connecting the assembly to the sandline. If the tool was dropped from surface without wireline, a spear point was fitted so that an overshot could be used to retrieve the instrument.

The instrument itself is contained within a protective casing. A heat shield must be used at high temperatures since the photographic film is heat-sensitive. The instrument can be run in several ways:

(a) lowered down and retrieved on wireline;

(b) dropped down and retrieved by tripping out the drill string at the end of a bit run;

(c) dropped down and retrieved by an overshot.

Non-magnetic Collars

In a magnetic single shot, the magnetic compass responds to the Earth's magnetic field. It is important, therefore, that any local magnetic field in the drill string is not allowed to distort the compass reading. The steel drill collars or drilling bit may become magnetized, creating "poles" especially near connections. To isolate the magnetic compass from possible distortion, the instrument must be contained within a non-magnetic environment. The instrument barrel and the drill collar within which it is measuring the survey angles must be made of non-magnetic materials. Non-magnetic drill collars (usually referred to as "monels") are made of alloys containing copper, nickel, chromium and other metals. The physical properties of the monels using these alloys are slightly different from regular steel collars. Monel collars should be handled carefully to prevent damage, because of their additional cost.

The number of non-magnetic collars required depends on several factors, including the direction and inclination of the hole, and the geographical location of the well as measured by its latitude (Fig. 7.3). The compass responds to the horizontal component of the Earth's magnetic field; in high northern latitudes this component is small. The compass may not distinguish between this small component and the other sources of magnetism nearby and so more monel collars should be run to minimize the effect of these other sources. Charts are available to estimate the number of monels that should be run and the spacing of the compass within the collars (Fig. 7.4).

Magnetic Single Shot

The instrument barrel of a magnetic single shot has several components, as shown in Fig. 7.5.

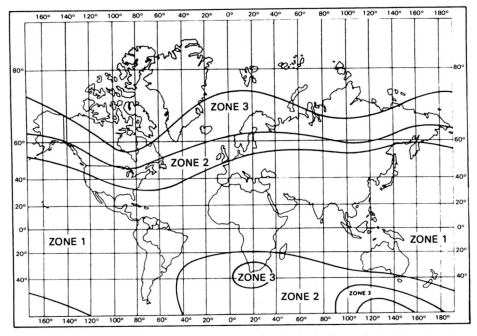

Fig. 7.3. Map showing different zones of magnetic field strength (courtesy of Eastman-Christensen).

(a) A sensitive angle measuring unit that indicates the direction and inclination of the hole.

(b) A film holder or "disc trap" that retains the small photographic disc. The disc must be inserted using a special loading device to prevent exposing the film to daylight.

(c) A camera unit and lamp assembly to photograph the position of the angle-indicating unit. The lights are switched on to allow the camera to capture an image of the pendulum superimposed onto the compass card. The camera is already focused so that the whole process is automatic.

(d) A timer or motion-sensing device that closes an electrical circuit and activates the camera at the correct moment. The timer can be set on surface to allow sufficient time for the tool to be run and landed in position before the camera operates. If the timer is set too early, the picture is taken while the tool is still moving. If the timer is set too late, time is wasted while waiting for the camera to operate. To obviate this problem, a motion sensor is installed that operates the camera when the instrument remains stationary for more than 30 seconds. Provided the tool does not get stuck for more than 30 seconds on the way down, the camera will operate when the tool is landed on the baffle plate.

105

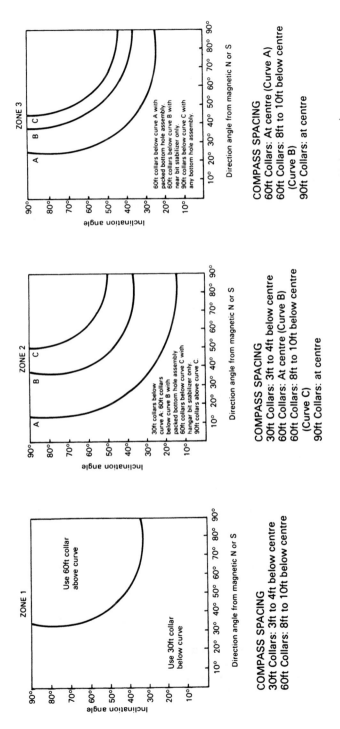

Fig. 7.4. Guide for numbers of non-magnetic drill collars and spacing of compass (courtesy of Eastman-Christensen).

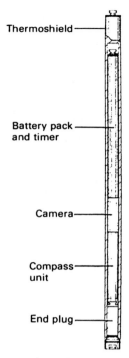

Fig. 7.5. Main components of single shot instrument (with heatshield) (courtesy of Eastman-Christensen).

(e) A battery pack that provides the power required to operate the camera, timer and lights.

The components of the angle-measuring unit are shown in Fig. 7.6. All the parts are sealed in a fluid chamber to provide a cushioning effect. The pendulum remains vertical when the instrument is lying at an angle. The distance between the cross-hair of the pendulum and the central axis of the tool gives a measure of the inclination. A glass plate with concentric rings provides a scale to allow direct reading of the inclination from the disc.

The magnetic compass is fixed to a compass card that aligns itself to Magnetic North. In high northern latitudes, where the Earth's magnetic field dips steeply, the card must be kept level by a compensating mechanism. When the cross-hair of the pendulum is photographed on top of the compass card, the hole direction can be read by interpolating between the radial lines. Depending on the expected inclination of the hole there are different sizes of angle units with scales of 0–10°, 0–20° or 15–90°. The appropriate angle unit should be chosen to make it easier to read off the results. On surface the disc is extracted from the instrument barrel, developed and read. To prevent reading errors two people should check

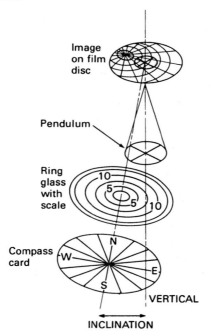

Image
on film
disc

Pendulum

Ring
glass
with
scale

Compass
card

N

W

E

S

VERTICAL

INCLINATION

Fig. 7.6. Diagrammatic view of single shot instrument.

the pictures independently. Examples of developed single shot pictures are shown in Fig. 7.7. To read off the angles, a straight line should be drawn from the centre of the ring glass through the cross-hair of the pendulum. The inclination is found by counting rings outwards from the centre to the cross-hair. The direction is found by interpolating between the radial lines around the perimeter of the disc.

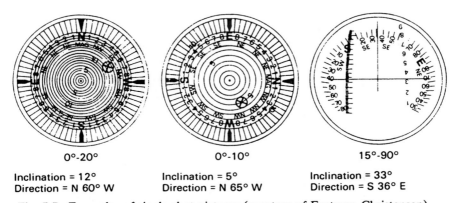

0°-20°	0°-10°	15°-90°
Inclination = 12°	Inclination = 5°	Inclination = 33°
Direction = N 60° W	Direction = N 65° W	Direction = S 36° E

Fig. 7.7. Examples of single shot pictures (courtesy of Eastman-Christensen).

The results from a magnetic single shot picture must be corrected for magnetic declination. This is the difference in angle between Magnetic North and True North. All survey results should be reported as true bearings. The amount of declination depends on geographical location. In the North Sea area Magnetic North is about 7° west of True North (Fig. 7.8). A single shot reading of N 60° E therefore becomes a true bearing of N 53° E. If the same survey were taken in California, where the magnetic declination is 15° East, the final result would be N 75° E. Due to the changing position of the magnetic poles there is a slight variation in magnetic declination with time. This variation may be 0–30 minutes per year, depending on geographical location. Notice also that the survey depth recorded for each survey station is the depth at which the instrument took the survey. The distance between the instrument and the bottom of the hole or bit must therefore be known in order to find the survey depth.

The normal procedure is to take a single shot survey after making a connection with the bit off-bottom. The instrument is checked, the timer is set and the tool is lowered down the drill string on the wireline. In highly deviated wells, sinker bars will be necessary to help the instrument travel down through the mud to reach the baffle plate. In some cases the instrument may have to be pumped down. The drill pipe should be reciprocated while the tool is being run, to prevent sticking. The running speed should be slowed as the instrument approaches the baffle plate. The drill pipe should not be moved during the time when the instrument is taking the survey. On floating rigs it may be necessary to put the bit on bottom to prevent pipe movement. After waiting for the camera to operate, one can pull the tool out of the hole again while reciprocating the drill string. Once the instrument barrel is recovered, the disc is removed and immersed in developing fluid. The developed picture is then read by the surveyor.

Any malfunction of the tool may not be discovered until the picture is developed. Several faults may occur.

(a) The timer may have been incorrectly set so that the survey was not taken at the correct depth.

For North Sea magnetic declination = -7
N60E (magnetic) = N53E (true)

For California magnetic declination = +15
N60E (magnetic) = N75E (true)

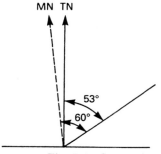
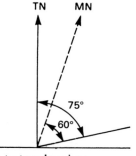

Fig. 7.8. Converting magnetic bearings to true bearings.

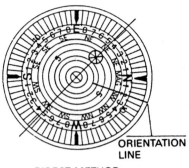

ORIENTATION
LINE
DIRECT METHOD
TOOL FACE N20°W
HOLE DIRECTION N45°E

Fig. 7.9. Toolface orientation from single shot (courtesy of Eastman-Christensen).

(b) If the lights have not operated correctly, there will be no picture.
(c) Damaged batteries may cause a complete power failure.
(d) Excessive vibration or hitting some obstruction may damage the angle unit itself.

To avoid the problem of a mis-run, tandem instruments can be run in the same assembly. Tandem instruments are also useful for checking reliability and accuracy of the results. The accuracy of a magnetic single shot is generally about ±0.3° for inclination and ±2° for azimuth. This is normally acceptable for plotting the course of the well while the hole is being drilled, and for orienting deflection tools. For orientation of the toolface, a reference mark is made on the picture showing the direction of the toolface heading (Fig. 7.9).

A magnetic surveying instrument cannot give accurate results in cased hole or near other wells that have been cased. The other major disadvantage of the single shot is the time taken to run and retrieve the instrument, which may be ½–1½ hours depending on the depth. If the drill string is allowed to remain stationary in a directional well, there is the risk of differential sticking. To prevent this the hole should be circulated to condition the mud prior to running the survey. The drill string should never remain static, except when the survey tool is actually on bottom taking the survey.

Magnetic Multishot

The magnetic multishot instrument works on the same principle as the single shot but is capable of taking a series of pictures at pre-set intervals. By dropping the multishot down the drill string just prior to pulling out of the hole, it is possible to survey the entire well path while tripping out. As with the single shot, the surveys must be taken inside the non-magnetic drill collars. The results of the multishot survey can then be compared with the individual single shot surveys taken as the well was being drilled. It is common practice to drop a multishot when a casing point is reached (Fig. 7.10).

The angle-measuring unit is essentially the same as in the single shot. The multishot however contains a reel of photographic film that is moved on by a mechanism controlled by a timer. The timer is set to advance the

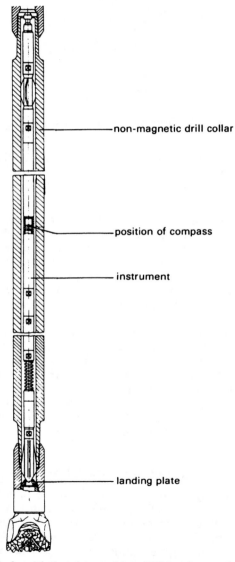

non-magnetic drill collar

position of compass

instrument

landing plate

Fig. 7.10. Position of multi-shot dropped into BHA prior to tripping out (courtesy of Eastman-Christensen).

film and take photographs at selected time intervals (e.g. every 30 seconds). Before dropping the multishot, the surveyor synchronizes his own watch with the timer so that he knows when the surveys are being taken. As each stand of drill pipe is being tripped out the surveyor must relate the survey depth to each picture. The pipe will be stationary long enough at each connection to allow two surveys, which should be identical. Those surveys taken between connections when the pipe is moving will be ignored.

When the instrument is recovered from the drill string at the surface, the film is removed and developed. The results will give the trajectory of that section of the hole at 90-ft intervals. The multishot results are usually preferred to the single shot results, since the survey interval is generally closer and the results are obtained in a single operation.

More recently, electronic multishots have been introduced in which solid state sensors are used to measure the angles and the results are stored in the tool's memory. Once the tool is recovered, the surveys can be obtained by linking the tool to a surface computer. This kind of instrument eliminates any reading errors that might occur with the older type of multishot. It also gives a higher accuracy ($\pm0.2°$ for inclination, $\pm1°$ for azimuth).

Gyro Single Shot

When surveying in cased hole with a magnetic compass, the presence of the steel casing will give erroneous results. This is also true when surveying in open hole when there are cased wellbores nearby. When kicking off a directional well from a multiwell platform, a magnetic single shot will be unreliable owing to the close proximity of adjacent wells. Under these circumstances the magnetic compass must be replaced by a gyroscopic compass that will not be affected by the presence of magnetic fields.

The gyroscope consists of a spinning wheel mounted on a horizontal axis and driven by an electric motor at speeds up to 40,000 rpm. The direction in which the gyro is spinning is maintained by its own inertia, and so can be used as a reference for measuring azimuth (Fig. 7.11). An outer and inner gimbal arrangement allows the gyro to maintain its pre-determined direction regardless of how the instrument is positioned in a deviated wellbore. Figure 7.12 shows the internal components of a gyro.

Before running the gyro single shot, the gyro must be aligned with a known reference direction, which is usually True North. This is done by accurate sighting of a benchmark using a telescope. With the gyro set on this heading the instrument is mounted into the survey tool assembly and lowered down the drill string on wireline. Inside the instrument the gyro is connected to a compass card. When the survey is taken the camera takes a photograph of a pendulum superimposed on the compass just as in the magnetic single shot. However, when the survey result is read from the picture the direction is referenced to True North and no correction for magnetic declination is applied.

One major problem with gyro surveys is to make allowance for gyro drift. As the tool is being run in the hole, the orientation of the gyro may

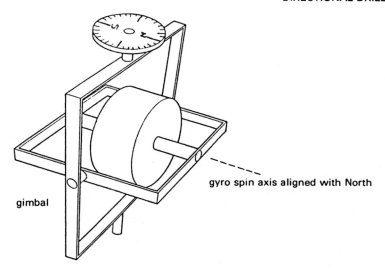

gyro spin axis aligned with North

gimbal

Fig. 7.11. Principle of gyro instrument.

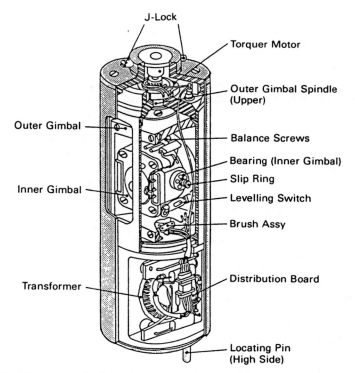

J-Lock

Torquer Motor

Outer Gimbal Spindle
(Upper)

Outer Gimbal

Balance Screws

Bearing (Inner Gimbal)

Slip Ring

Inner Gimbal

Levelling Switch

Brush Assy

Transformer

Distribution Board

Locating Pin
(High Side)

Fig. 7.12. Diagrammatic view of gyroscope (courtesy of AMF Scientific Drilling).

drift slightly from its original heading owing to the effect of unbalanced forces acting on it. After the tool is recovered from the hole, the alignment of the gyro should be checked. A correction is then applied to the survey result. Unlike a magnetic single shot, a gyro must be run on wireline, since the sensitive mechanism could easily be damaged. The major application for a gyro single shot is for surveying or orienting deflecting tools near steel casing.

Gyro Multishot

Once a string of casing has been run in the hole, the definitive trajectory of the cased borehole is usually provided by a gyro multishot survey. The gyro multishot is run on wireline and the surveys are taken while running into the hole. This is to reduce the error caused by gyro drift, which becomes significant over longer times. Gyro drift does not increase uniformly with time. To correct the survey results for the effect of gyro drift a series of drift checks are made both running in and coming out of the hole. The gyro is held stationary for a few minutes, allowing a number of pictures to be taken at the same point. A drift correction chart can then be drawn up to adjust the raw survey results.

STEERING TOOLS

During the critical kick-off stage in the drilling of a directional well it is necessary to survey the well at close intervals. If the well path is not following the intended trajectory, the bent sub must be re-oriented to bring the well back on course. Frequent surveying and re-orientation using conventional single shot instruments is very time-consuming and expensive on offshore platforms.

One solution to this problem is to have an instrument that can be run in the bottom hole assembly to survey the well continuously while it is being drilled with a downhole motor. These surveying tools are generally called "steering tools", since they provide the directional driller with the necessary information to steer the bit in the correct direction. The tool consists of an electronic probe that is run into the hole on a conductor line. The probe is seated in the orienting sub just above the bent sub. Within the probe are the electronic sensors that measure hole inclination, azimuth and toolface. The survey results are transmitted from the probe via the conductor line to surface, where a computer analyses the signal and gives a digital display of the angles measured (Fig. 7.13).

This method of surveying offers several advantages over single shots.

(a) Rig time is saved by eliminating the large number of wireline trips required to take surveys and to check orientation.
(b) Continuous monitoring will reduce the risk of the well straying off course, and therefore reduce the number of correction runs.
(c) Owing to better control the well path should be smoother, with fewer dog-legs.

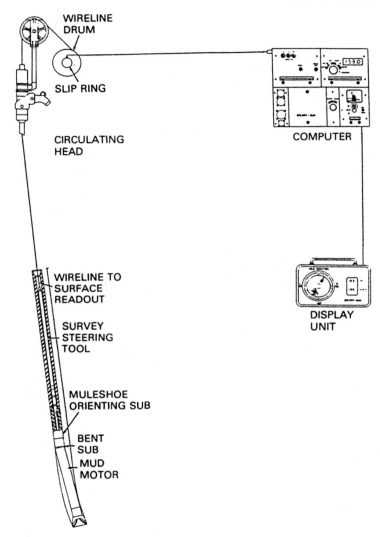

Fig. 7.13. Components of a steering tool system (courtesy of NL Sperry Sun).

(d) The toolface heading can be monitored during drilling to account for reactive torque.

Downhole Tools

Since the probe is effectively part of the BHA, the sensors must be rugged enough to withstand the vibration and shock loads generated downhole. Gyroscope devices are too sensitive to be used for this application. Surface readout gyros, however, can be used to give continuous monitoring of toolface while carrying out orientation. These tools are used when kicking off near casing, where magnetic interference is present. Once the bent sub is correctly oriented, however, the gyro must be pulled out, otherwise its sensitive mechanism would be damaged as drilling begins.

The sensors used in steering tools and MWD tools are solid-state electronic devices known as magnetometers and accelerometers which respond to the earth's magnetic field and gravitational field respectively. Since the magnetometers may be affected by the steel collars and drill pipe, the probe must be seated within non-magnetic collars. The probe slots into the muleshoe key, which is aligned with the scribe line of the bent sub. The probe therefore measures the direction in which the scribe line is pointing. The orientation of the bent sub can be measured relative to Magnetic North (magnetic toolface) or with respect to the High Side of the hole (gravity toolface). At low inclinations (0–8°) magnetic toolface is used, since the High Side is not well-defined at that stage. Once the angle increases, however, and the hole direction becomes established, the gravity toolface is used (i.e. toolface is reported as a number of degrees to the left or right of High Side). The High Side of the hole can be defined by the accelerometers. The High Side is directly opposite to the gravity vector, which is the sum of the three gravitational components measured by the accelerometers.

Surface Equipment

Since rotation of the drill string is eliminated by the use of a downhole motor driving the bit, the kelly can be replaced by a circulating head. This allows mud to be pumped down the drill pipe while the wireline is in the hole. The wireline passes through a pack-off or sealing device above the circulating head (Fig. 7.13). When a new section of pipe is being added the wireline probe must first be tripped out of the hole, and then run in again after the connection has been made. To save time, connections are made every three joints (i.e. the circulating head is installed on top of a stand of pipe so that 90 ft can be drilled at a time without tripping the probe).

Another device that can save time is the side-entry sub. This allows the wireline to pass from the drill string out to the annulus at some point beneath the rig floor. Since the wireline no longer interferes with the connection, the probe need not be pulled out (Fig. 7.14). Special care must be taken when setting the slips, however, since the wireline may be damaged. The electrical signals sent from the probe through the conductor line are decoded by a surface computer. The results are relayed to a rig-floor display so that the directional driller can get continuous updates of the measured angles. By observing the dial, the directional driller can see at a glance how the toolface is changing as drilling proceeds.

The disadvantages of steering tools are that:

(a) they can only be used with downhole motors, since drill string rotation cannot be permitted;

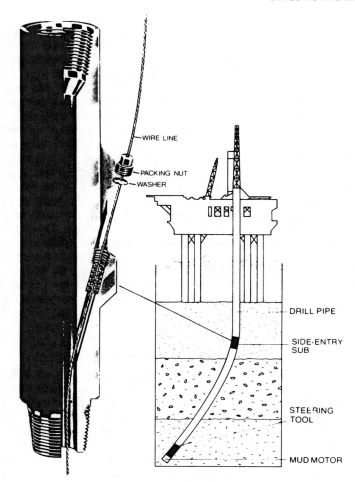

Fig. 7.14. Side-entry sub (courtesy of NL Sperry Sun).

(b) since they make use of magnetometers they will not give reliable
    survey results near casing;
(c) unless a side-entry sub is used the probe must be tripped out when
    connections are to be 90 ft drilled.

SOLID STATE DIRECTIONAL SENSORS

To survey the borehole continuously while drilling requires the use of
rugged, solid-state accelerometers and magnetometers. Accelerometers
measure components of the Earth's gravitational field, while magnetometers measure components of the Earth's magnetic field. In each case the

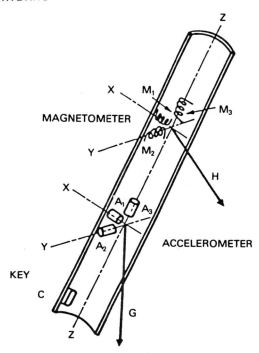

Fig. 7.15. Arrangement of accelerometers and magnetometers along three orthogonal axes (courtesy of Eastman-Christensen).

fields are acting in a specific direction, and by measuring the orientation of the surveying tool with respect to that direction the inclination, azimuth and toolface can be determined (see Fig. 7.15).

Accelerometers

Figure 7.16 illustrates the principle on which an accelerometer works. A test mass is held in position by a quartz hinge that allows the mass to move along one axis only. When it is positioned in the borehole the component of gravity acting along that axis will tend to move the mass. Movement of the test mass creates an imbalance between the capacitors, which is detected by the servoamplifier. An electric current is then sent through the coil, creating an opposing force that tends to restore the original position of the test mass. The greater the gravitational component, the greater the current required to oppose it. The voltage drop over a known resistance therefore gives a direct measurement of the gravitational component.

For a triaxial accelerometer, the vector sum of the three components must be equal to g, the acceleration due to gravity. Since g can be determined at any location by other means, only two accelerometers are actually required. However, a triaxial accelerometer does provide a means of checking the outputs and identifying any errors.

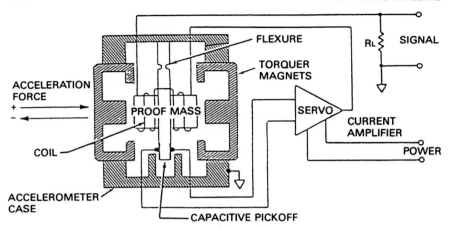

Fig. 7.16. Operation of accelerometer (courtesy of Sundstrand Data Control Inc.).

Magnetometers

A magnetometer is a sensor that detects and measures the strength of the Earth's magnetic field along a fixed axis. Figure 7.17 shows a soft iron core with a coil of wire wrapped around it. If this core (or toroid) is positioned within an alternating magnetic field, a magnetic flux will be concentrated within the toroid and a current will be generated in the wire. The size of the current will depend on the amount of permeable material that is exposed to the magnetic field. If the toroid is placed at 90° to the field lines, the current will be large. As the toroid is rotated such that less area is exposed to the field, the current will reduce. The size of the current picked up in the coil can therefore be used as a means of measuring the angle between the magnetic field and the coil. This current, however, will only be produced if the magnetic field is alternating. The Earth's magnetic field is constant. Nor is it feasible to move the toroid to generate a current, since this would reduce the accuracy of the directional measurements. The toroid must remain fixed and aligned with one of the reference axes of the surveying tool. By using a 'flux gate' device the strength of the magnetic field along that particular axis can be measured.

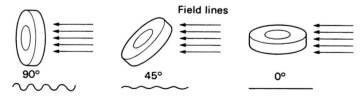

Fig. 7.17. Current generated by toroid in alternating field.

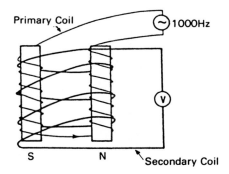

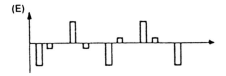

Fig. 7.18. Flux gate magnetometer. The peak of the resultant voltage (E) is proportional to the strength of the field parallel to the axis of the coils.

To illustrate the principle of a flux gate magnetometer, consider two identical cores as shown in Fig. 7.18. The cores are of equally high permeability and have primary and secondary coils of wire that are wound in opposite directions. An alternating current is passed through the primary coil, generating a magnetic field, so that the cores become saturated. In the absence of any external field the combined voltage output will be zero, since the coils are wound in opposite directions. However if there is an external magnetic field it will cause saturation in one coil before the other. The output voltages will then be out of phase, causing voltage pulses. The pick-up coil placed at some angle to the external field will therefore produce a voltage that is related to the rate of change of flux through the toroid. The size of the voltage is thus related to the strength of the external field. A triaxial flux gate magnetometer can therefore measure components of the Earth's magnetic field along three orthogonal axes.

Derivation of Inclination, Azimuth and Toolface

The voltage outputs from the accelerometers are denoted by G_x, G_y and G_z, corresponding to the three orthogonal axes (Fig. 7.15). Similarly the magnetometer outputs are H_x, H_y and H_z. Notice that the z axis points down the axis of the tool and the y axis is defined as being in line with the toolface. This can be used as a reference axis that can be related to the scribe line of the bent sub.

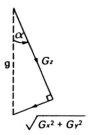

Fig. 7.19. Inclination angle: $\alpha = [(G_x^2 + G_y^2)/G_z]^{1/2}$.

Inclination

Looking at right angles to the vertical plane, the inclination is the angle measured from vertical to the axis of the z accelerometer (Fig. 7.19). The inclination α can be found from

$$\tan \alpha = \frac{(G_x^2 + G_y^2)^{1/2}}{G_z}$$

Toolface

Gravity toolface (GTF) is the angle looking downhole between the High side of the hole, as defined by the gravity vector and the y accelerometer (Fig. 7.20). This angle can be found from:

$$\tan \text{GTF} = \frac{G_x}{G_y}$$

Magnetic toolface (MTF) is the angle between Magnetic North and the y axis, such that:

$$\tan \text{MTF} = \frac{H_x}{H_y}$$

Azimuth

This is the angle measured clockwise from North to the position of the z axis as seen in the horizontal plane. To calculate the azimuth, the magnetometer and accelerometer readings must be resolved into two axes

Fig. 7.20. Toolface angle: GTF = $\tan^{-1}(G_x/G_y)$.

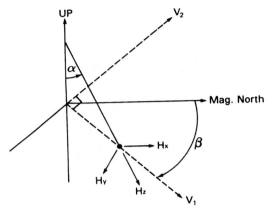

Fig. 7.21. Azimuth angle:

$$\beta = \tan^{-1}\left(\frac{g(H_xG_y - H_yG_x)}{H_z(G_x^2 + G_y^2) + G_z(H_yG_y + H_xG_x)}\right)$$

(Fig. 7.21). The V_1 axis is the projection of the hole direction onto the horizontal plane. The V_2 axis is at right angles to the V_1 axis such that the azimuth β can be found from

$$\tan \beta = \frac{\text{Resultant } V_2 \text{ component}}{\text{Resultant } V_1 \text{ component}}$$

Resolving H_x, H_y and H_z in terms of V_1 and V_2, the following equations are obtained:

$$V_1 = H_z \sin \alpha + H_y \cos \text{TF} \cos \alpha + H_x \sin \text{TF} \cos \alpha$$
$$V_2 = H_x \cos \text{TF} - H_y \sin \text{TF}$$

Substituting from the previous relationships,

$$\sin \alpha = \frac{(G_x^2 + G_y^2)^{1/2}}{g_z}$$

$$\cos \alpha = \frac{G_z}{g}$$

$$\sin \text{TF} = \frac{G_x}{(G_x^2 + G_y^2)^{1/2}}$$

$$\cos \text{TF} = \frac{G_y}{(G_x^2 + G_y^2)^{1/2}}$$

The final expression for azimuth is

$$\beta = \tan^{-1}\left(\frac{V_2}{V_1}\right) = \tan^{-1}\left(\frac{g(H_xG_y - H_yG_x)}{H_z(G_x^2 + G_y^2) + G_z(H_yG_y + H_xG_x)}\right)$$

where $g = (G_x^2 + G_y^2 + G_z^2)^{1/2}$. Notice that the expression for azimuth contains both accelerometer and magnetometer results.

RATE GYROS

In conventional gyros the spin axis must be carefully aligned with True North on surface before being run into the hole. Since all subsequent surveys are referenced to this direction, any misalignment on surface introduces systematic errors in the survey results. While the instrument is being run in the hole, the tendency for the gyro to drift away from its original heading also introduces errors. Although the amount of drift can be measured periodically during the run and corrections can be applied to the results, the drift error cannot be completely eliminated.

In areas where there is a need for high-accuracy surveys, a new genera-

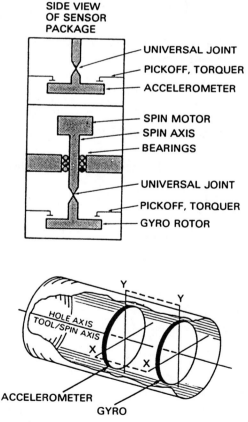

Fig. 7.22. Rate gyro (after Uttecht and De Wardt; courtesy of SPE).

tion of gyroscope instruments is now being employed. These tools are based on the principle of rate gyrocompassing and are usually referred to as "rate gyros". The advantage of these instruments is that they do not need to be aligned with True North on surface. The rate gyro senses True North independently at each survey station, and so is not subject to the cumulative effects of conventional gyro drift. This North-seeking capability removes many of the errors inherent in conventional gyro surveying.

The basic components of the sensor package are shown in Fig. 7.22. The gyroscope and accelerometer are both aligned with the tool axis. The spinning gyro is not free to change its orientation with respect to the instrument case as it is in conventional gyros. A sensitive torque-balancing system maintains the alignment of the gyro with the axis of the tool. The instrument will be stationary in the borehole while taking a survey. The only force causing movement of the case is the Earth's rotation. The output torque required to maintain alignment is directly related to the rate of angular movement of the Earth. With knowledge of the Earth's latitude at this point, the measurement can be resolved into horizontal and vertical components. The horizontal component by definition always points True North, thereby providing a reference direction. At the same time, the accelerometer measures components of the earth's gravitational force, from which the hole inclination can be determined. The azimuth of the wellbore can be found by combining the gyro and accelerometer outputs.

The accuracy of the tool is further improved by taking a series of readings at each survey station and averaging the results. The sensor package can be rotated and the survey repeated to eliminate any error due to instrument bias.

Running Procedure

Each tool must be calibrated on a test stand before being sent to the rig. The instrument is oriented to the direction and inclination likely to be encountered downhole. A series of calibration factors can be derived from the results of the tool's response on the test stand. These calibration factors are later used at the well-site to calculate the final survey results. After warming up, the probe is run down the hole on conductor line at a speed of 300–500 ft/min. Surveys can be run while running in or pulling out. To take a survey the instrument is held stationary for a short period of about 1 min. This allows the sensors to make the correct measurements. The signals are sent from the probe via the conductor line to the surface computer, which outputs the inclination and azimuth.

The major application of these tools is to obtain a definitive survey of the well after casing has been run. The downhole tool is designed to be fitted inside a small-diameter pressure housing. This allows the tool to be run through drill pipe, casing or production tubing. The survey can be taken faster than a conventional gyro, since no drift checks are necessary.

The accuracy of the rate gyro is generally given as $\pm0.5°$ in azimuth and $\pm0.05°$ in inclination for inclinations greater than 5°. These instruments can survey the borehole to a lateral displacement error of 3 ft/1000 ft. The accuracy is dependent on hole angle and latitude. At high inclinations the accelerometer readings become unreliable owing to the small axial component of gravity. The North-seeking capability of the tool is reduced at latitudes over 70° from the equator, since the horizontal component of the Earth's spin vector becomes very small.

GCT (Schlumberger)

The GCT, or continuous guidance tool, was introduced in 1981 as a means of surveying the cased borehole without having to stop at each survey station. This allows an accurate survey of the complete trajectory of the well in a much shorter time, although the alignment procedure at the well-site may take 1 hour.

The operation of the tool is based on an inertial platform consisting of a dual-axis gyroscope that is North-seeking and a dual-axis accelerometer. The gyro spin axis is horizontal and parallel with one axis of the accelerometer. Both sensors are mounted on a gimbal system (Fig. 7.23). During calibrations at the well site the gyro is oriented towards True North and the spin axis is checked to ensure it is horizontal. When the tool is run

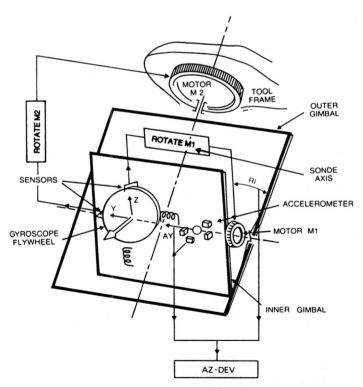

Fig. 7.23. Continuous guidance tool (courtesy of Schlumberger).

in the hole, the position of tool in the hole does not affect the gyro orientation. Azimuth and inclination are calculated from the A_y accelerometer reading and the angle between the inner and outer gimbals (R_1).

As the tool is being run continuously, readings are averaged over 10-sec periods. At a running speed of up to 10,000 ft/h, this gives a survey point every 25–30 ft. Surveys are actually taken both while running in and pulling out of the hole. Owing to the small OD ($3\frac{5}{8}$ in.) it can be run to the bottom of the hole even through 7 inch liners. A surface computer processes the data and determines the azimuth and inclination.

The accuracy of the GCT is again dependent on latitude and inclination. The error in terms of lateral displacement is given by

$$E = \frac{0.4}{100}\left(\frac{\cos 45°}{\cos \text{LAT}}\right)H + \left(\frac{0.06}{100}\right)L$$

where L = length of the hole (measured depth)
 LAT = latitude (degrees)
 H = horizontal displacement (ft).

INERTIAL NAVIGATION (FINDS)

On multiwell platforms the close proximity of wells around kick-off depth requires that the position of each wellbore is known to a high degree of accuracy. This is necessary to prevent intersections beneath the platform and to avoid having to close-in producing wells while drilling proceeds. Conventional gyros were not considered sufficiently accurate for this application in near-vertical wells and so alternative methods of surveying were investigated. In 1979 a new survey tool was introduced known as FINDS (Ferranti Inertial Navigation Directional Surveyor). This tool operates on the same navigation principles as used in aircraft and missile guidance systems. It is based on an inertial platform that contains three accelerometers and three gyroscopes set on orthogonal axes (East/West, North/South and vertical as shown in Fig. 7.24). The inertial platform is mounted on gimbals that are driven by a servomechanism. On surface, the platform is levelled and aligned approximately with True North. The gyroscopes then orient the platform automatically so that the N/S accelerometer is pointing directly at True North. Any tilt or misalignment of the platform during the surveying operations is detected by the gyros and corrected by the servomechanism.

Since the accelerometers are oriented exactly along the reference axes, any acceleration of the system can be measured and resolved in three directions. The normal surveying procedure is to run the tool for 1 min. (transit time) and then hold it steady for 1 min. (fix time). Each accelerometer reading is integrated to provide velocity components, which are stored in the tool's downhole memory. This procedure is repeated until the tool reaches the casing shoe. When the tool is pulled back to surface,

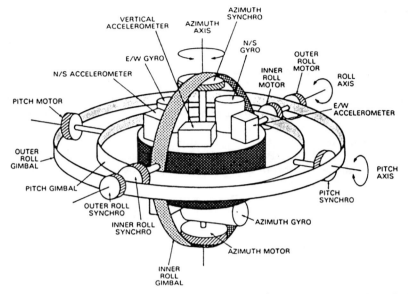

Fig. 7.24. Inertial navigation tool (after Morgan and Howe; courtesy of the SPE).

the recorded data are integrated again by the surface computer to determine the displacements ΔN, ΔE and ΔV. If one knows the surface coordinates of the well, these incremental displacements can be added successively to determine the trajectory of the wellbore. The inclination and azimuth of the wellpath are not measured directly by the FINDS tool, but they can be back-calculated if necessary from the coordinates.

One of the major disadvantages of the FINDS tool is the large diameter (10⅜ in. OD), which limits its use to surveying in larger-diameter casing. In the North Sea it is generally used to survey down to the 13⅜ in. casing shoe. It is over this upper section of hole that survey accuracy is most important to avoid collisions. Its accuracy is reported as ±0.5 ft/1000 ft of measured depth. It is generally considered to be the most accurate tool presently available for borehole surveying.

SURVEYING TECHNIQUES TO LOCATE BLOW-OUTS

When a blow-out occurs during drilling operations the resulting damage to the rig and surface equipment may make a re-entry too dangerous or impossible. In this case a relief well must be directionally drilled from some safe distance away to intersect the wild well. Heavy mud or cement can then be pumped down the relief well to control the bottom hole pressure. Very close directional control is required in drilling the relief well in order

to hit the target, otherwise the kill operation will not be successful. The job
is sometimes made more difficult by the lack of survey data on the wild
well, so that the bottom hole location is not known very accurately.

Several relief wells have been drilled successfully by monitoring the
relative distance between the two wells by means of logging tools. There
are basically two types of logging tools that can detect the presence of steel
casing in an adjacent well.

(a) Resistivity logs are commonly used in open hole formation evaluation
    to measure the ability of the formation to conduct electrical current. If
    a steel casing is present in the formation, the resistivity tool will detect
    this as an anomaly. By comparing the expected response of the
    formation itself and the actual response due to the casing, the inter-
    well distance can be estimated. In general, a resistivity tool can detect
    the presence of a casing string up to 100 ft away. However, the tool
    gives no indication of the direction in which the casing lies.

(b) Magnetic logs will provide both distance and direction to the casing of
    the wild well. The current emitted by the logging tool creates a
    magnetic field around the casing. This magnetic field is detected by
    magnetometers inside the logging tool. The effect of the Earth's
    magnetic field is first subtracted from the tool's response, leaving the
    net effect solely due to the casing. A vector analysis of the magneto-
    meter readings gives the direction of the source, and the inter-well
    distance. With this information the trajectory of the relief can be
    planned to intersect the target. The magnetic tool must be within about
    50 ft to detect the casing.

A summary of the accuracy figures for the more commonly used surveying
tools is given in Table 7.1.

TABLE 7.1 Comparison of Different Types of Survey Instrument

Type of survey instrument	Estimated lateral error at 10,000 ft MD	Outside diameter	Key limitations
Conventional or free gyro	100 ft at $\alpha = 45°$	2–3 in.	Operates at $\alpha < 70°$ Alignment errors Drift errors
Magnetic	200 ft at $\alpha = 45°$	2.25 in.	Operates at $\alpha < 70°$ Sensitive to magnetic effects Correction for declination
Gyro compass	20–30 ft at $\alpha = 45°$	1.75–3 in.	Operates at $\alpha < 70°$
Inertial navitation	1 ft	10.6 in.	Complex mechanism Depth limitation

(After Brzezowski and Fagan.)

QUESTIONS

7.1. The two diagrams below show the results of magnetic single shot surveys. Read off the inclination and direction in each case.

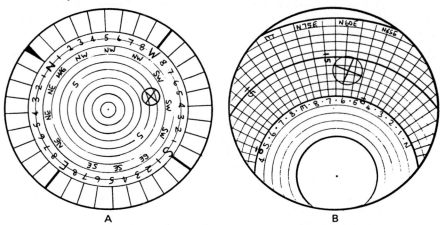

A   B

7.2. For each of the surveys in Question 7.1, correct the azimuth by applying the magnetic declination in the following areas:
(i) Gulf of Mexico, declination = 7° East.
(ii) Offshore Canada, declination = 26° West.

7.3. An operator is planning to drill a directional well in the North Sea. The inclination and direction will be approximately 40° and N 20° W respectively. Determine:
(a) the number of non-magnetic collars required;
(b) the compass spacing within the collars.
If the well plan is changed such that the direction is now N 60° W, will this affect the number of collars required?

7.4. For a well whose inclination is 50° and whose direction is N 60° E, determine the number of non-magnetic drill collars required in the following areas:
(a) Saudi Arabia;
(b) Alaska.
Explain why more collars are required in more Northerly latitudes.

7.5. List the major applications of gyro surveying instruments.

7.6. The GCT survey tool is being used in a directional well whose total measured depth is 12,000 ft and horizontal displacement is 3500 ft. If the latitude is 60° N, calculate the accuracy in terms of lateral displacement.

FURTHER READING

"Hand-held calculator assists in directional drilling control", J. L. Marsh, *P.E.I.*, July, September 1982.

"Analysis of alternate borehole survey systems", S. Brzezowski and J. Fagan, 39th Annual Meeting, Institute of Navigation, June 1983.

"How to avoid gyro mis-runs", J. Plite and S. Blount, *Oil and Gas Journal*, 14 January 1980.

"Calculation of NMDC length required for various latitudes developed from field measurements of drill string magnetisation", S. J. Grindrod and J. M. Wolff, I.A.D.C./S.P.E. paper no. 11382.

"Rate gyro surveying of wellbores in the Rocky Mountains", J. Wright, S.P.E. paper no. 11841.

"Application of small diameter, inertial grade gyroscopes significantly reduces borehole position uncertainty", G. W. Uttecht and J. P. deWardt, I.A.D.C./S.P.E. paper no. 11358.

"Gyroscopes: theory and design", P. H. Savet, 1961.

"A new continuous guidance tool used for high accuracy directional surveys", D. J. Camden, J. E. Gartner and P. A. Moulin, S.P.E. paper no. 10057.

"A new generation directional survey system using continuous gyrocompassing techniques", A. C. Scott and J. W. Wright, S.P.E. paper no. 11169.

"Directional survey and proximity log analysis of a downhole well intersection", T. M. Warren, S.P.E. paper no. 10055.

"Directional surveying using inertial techniques—field experience in the northern North Sea", J. van Nispen and N. A. Howe, S.P.E. paper no. 10384.

"High accuracy directional surveying of wells employing inertial techniques", D. G. Morgan and A. Scott, O.T.C. paper no. 3359.

"How to specify and implement well surveys", J. L. Thorogood, *World Oil*, July 1986.

Chapter 8

MEASUREMENT WHILE DRILLING (MWD)

MWD is the process by which certain information is measured near the bit and transmitted to surface without interrupting normal drilling operations. The type of information may be:

(a) directional data (inclination, azimuth, toolface);
(b) formation characteristics (gamma-ray, resistivity logs);
(c) drilling parameters (downhole WOB, torque, rpm).

The sensors are installed in a special downhole tool made up as an integral part of the bottom hole assembly. Within the downhole tool there is also a transmitter to send the signals to surface via some kind of telemetry channel. The most common type of telemetry channel currently in use is the mud column inside the drill string. The signals are detected on surface, decoded and processed to provide the required information in a convenient and usable format. Figure 8.1 shows the main components of an MWD system. The great advantage of MWD is that it allows the driller and the geologist to effectively "see" what is happening downhole in real time. It therefore improves the decision-making process, since there is a delay of only a few minutes between measuring the parameters downhole and receiving the data on surface.

Although the concept of MWD is not new, it is only in recent years that advances in drilling technology have made MWD a reality. Electric logging introduced in the 1930s made a significant contribution towards identifying and evaluating formations. Its major disadvantage, however, was that the tool had to be run on wireline after the drill string had been pulled out of the hole. Furthermore, by the time the log was actually run, the effects of mud invasion prevented the measurement of the true characteristics of the formation. Since there was no definite means of identifying changes in lithology as the bit drilled through the different formations, important horizons were not detected. Subsequent logs sometimes showed that coring points at the top of the reservoir section had been missed, or that the bit

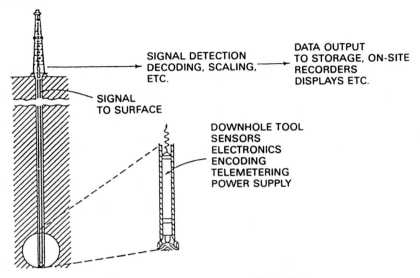

Fig. 8.1. Overview of measurement while drilling system.

had drilled too far into the water zone beneath the pay zone. Mudlogging and monitoring the ROP provided some indication of downhole conditions, but the time lag in waiting for cuttings to be circulated to surface made the process fairly inefficient. There was therefore a need for a system that provided instantaneous and continuous monitoring of the formation while drilling. The requirements of such a system were as follows:

(a) rugged and reliable sensors that could measure the required data at or near the bit under dynamic drilling conditions;
(b) a simple but effective method of transmitting the information to surface;
(c) a system that could be easily installed and operated on any rig without causing too much disruption to normal drilling practices;
(d) a system that was cost-effective and provided real benefits to the operator.

Several attempts were made at producing a system that satisfied these requirements. The major problem proved to be the telemetry link between downhole and surface. Between 1930 and 1960 four alternative telemetry systems had been investigated:

(a) electric conductors (hard-wire systems);
(b) electromagnetic radiation;
(c) seismic (acoustic) waves;
(d) mud pressure pulses.

Up to 1960 these telemetry systems were being investigated mainly for logging-while-drilling applications. The increasing use of directional drilling, especially in expensive offshore areas, provided an added incentive to develop an MWD system that could handle directional surveying data as well as formation evaluation data. It was soon realized that there was a greater commercial potential for directional MWD owing to the high cost of running conventional surveying tools on offshore platforms. In fact, the first MWD system to be commercially available in the North Sea (introduced in 1979) provided only directional data. This was closely followed by other tools that measured additional information such as drilling parameters and formation data. Only those MWD systems based on mud pressure pulse telemetry have so far proved technically and economically viable, although research is continuing into the other three telemetry methods.

Since 1979 there have been many improvements in MWD, offering more sensors and greater reliability. With more MWD companies entering the market, there is a wider variety of tools available and the competition has reduced prices. For directional drilling, MWD has become a standard technique in the North Sea. There is also much interest in using MWD for other applications, especially logging. In the development of multisensor tools, however, it has become clear that to provide real time information the data rate (the speed at which downhole data is sent to surface) must also be improved.

TELEMETRY CHANNELS

Hard-wire Methods

The most direct approach is to transmit electrical signals to surface via some kind of conductor. Hard-wire methods were first proposed in the 1930s as a means of sending formation data to surface while drilling. More recently two different methods have been under investigation:

Insulated conductor attached to drill pipe

This method employs a continuous conductor that forms an integral part of the drill pipe. Special connectors built into the tool joints provide conductivity throughout the length of the drill string. The sensors are located in a special drill collar. An armoured cable (or jumper) connects this collar with the lower end of the drill pipe, thus eliminating the need to have an integral circuit through the various components of the BHA. The length of the jumper must be equal to the length of the BHA to ensure correct tension is maintained. At the other end of the system there is an insulated slip ring mounted on top of the kelly. This is connected to the surface equipment that processes the signals and gives the final results (Fig. 8.2).

The major disadvantages of this system are:

(i)   the additional cost of a special string of drill pipe;
(ii) the difficulty in achieving a continuous circuit at connections.

Conductor cable running through drill string

To overcome the shortcomings of the first method, some companies have investigated the possibility of running an electrical wireline through the

(Drill-pipe conductor and
tool-joint connector)

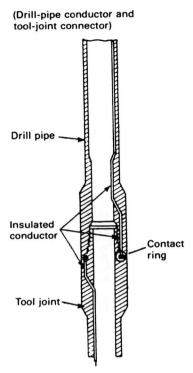

Drill pipe

Insulated
conductor

Contact
ring

Tool joint

Fig. 8.2. Hardwire system using special drill pipe with integral conductor.

drill string. The type of conductor is an armoured cable similar to that used for electrical wireline logging operations. The obvious disadvantage here is that the cable must be pulled out at connections as the hole gets deeper. This problem can be solved by storing extra length of cable on a spool inside the drill pipe. Electrical–mechanical latches built into the system allow the cable to be temporarily disconnected while the new joint is added. However, at the end of a bit run the entire length of cable must be retrieved first before tripping the drill string out of the hole (Fig. 8.3).

Although there are serious operational problems still to be overcome, hard wire systems do offer certain advantages over other telemetry methods.

(a) higher data rates allow more information to be transmitted in real time;
(b) no need to incorporate a downhole power source;
(c) two way communication is possible (i.e. signals could also be sent down to activate certain components such as an adjustable bent sub, or a downhole BOP);
(d) unlike some other methods there is effectively no depth limitation since signal attenuation is not a problem.

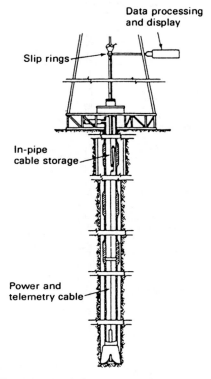

Fig. 8.3. Hardwire system using cable loop.

Electromagnetic Methods

Since the 1940s several companies have investigated the potential of sending signals through the Earth's crust by means of electromagnetic (radio) waves. A wave transmitter mounted within the BHA generates signals that can be modulated to send the required data in the form of a binary code. These signals are detected at surface by antenna placed on the ground around the rig site. This system offers certain advantages:

(a) no disruption to normal drilling operations;
(b) simpler rig-up than other methods;
(c) data can be transmitted while tripping.

These advantages, however, must be weighed against the problem of attenuation of the signals. Only low-frequency waves can be transmitted effectively, and these are sometimes difficult to distinguish from the frequencies emitted by electrical equipment on the rig (Fig. 8.4).

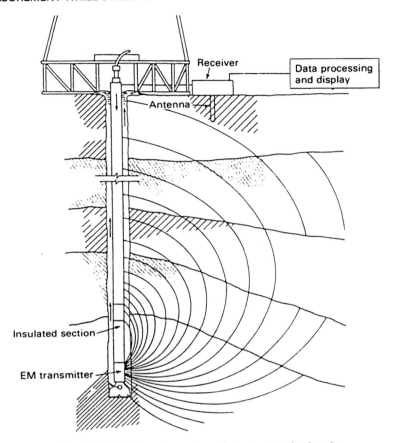

Fig. 8.4. MWD system using electromagnetic signals.

Acoustic Methods

The propagation of sound or seismic waves through the drill pipe provides another possible telemetry channel. To overcome background noise due to drilling operations, a large seismic generator would be required in the BHA. Attenuation along the drill string also makes it difficult to pick up the signal at surface. Repeater stations situated at intervals along the drill string can be used to amplify the signal but these introduce further complications and add to the cost of the system.

Mud Pulse Telemetry

All the MWD systems commercially available are based on some form of mud pulse telemetry. Although several different companies offer an MWD

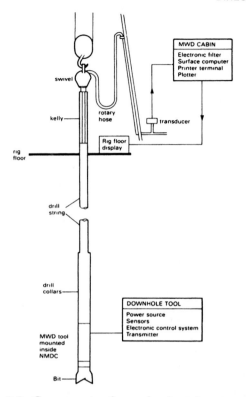

Fig. 8.5. Components of a mud pulse telemetry system.

service, there are certain similarities among the systems being used. The major components of a mud pulse telemetry system are shown in Fig. 8.5. The downhole components are all housed in a nonmagnetic drill collar. This is a special collar supplied by the MWD company. Its internal diameter is greater than the regular size since it has to accommodate the MWD tool components. The major components are:

(a) a power source to operate the tool;
(b) sensors to measure the required information;
(c) a transmitter to send the data to surface in the form of a code;
(d) a microprocessor or control system to coordinate the various functions of the tool;

The control system is designed to operate the tool when information is required (e.g. when a directional survey is to be taken). To initiate the survey, the downhole tool must recognize some physical change (e.g. when drill string rotation stops or when the pumps are shut off). From this point onwards the control system powers up the sensors, stores the information that has been measured and then activates the transmitter to send the data in the form of a coded message.

The surface equipment consists of:

(a) a standpipe pressure transducer to detect variations in pressure and convert these to electrical signals;
(b) an electronic filtering device to reduce or eliminate any interference from rig pumps or downhole motors that may also cause pressure variations;
(c) a surface computer to interpret the results;
(d) a rig-floor display to communicate the results to the driller, or plotting devices to produce continuous logs

The major advantage of mud pulse telemetry over other methods is its relative simplicity. No special drill pipe is required, there are no complications due to wireline in the hole, and only fairly minor alterations to normal drilling practices are necessary. The pressure pulses travel through the mud column at around 4000–5000 ft/s, but there are limits to the amount of information that can be sent in real time. Several companies are investigating methods of improving the data rate of existing MWD systems.

TRANSMISSION SYSTEMS

The major differences between existing mud pulse telemetry systems are due to the process by which the data is transmitted. There are three different methods currently being used to encode the data downhole and to decode the data on surface.

Positive Pulse System

Within the downhole tool (Fig. 8.6) there is a restrictor valve that is operated by a hydraulic actuator. When the valve is operated, it forms a temporary constriction in the flow of mud through the drill string which causes an increase in the standpipe pressure. To transmit data to surface, this valve is operated several times, creating a series of pulses that are detected by the transducer, and decoded by the surface computer. The surface computer initially recognizes a set of reference pulses, which are followed by the data pulses. The message is decoded by detecting the presence or absence of a pulse within a particular time-frame. This binary code can then be translated into a decimal result. A chart recorder is used to monitor the sequence of pulses. In the event of an electrical failure in the decoding mechanism, the pulse sequence can be decoded manually from the chart recorder by the service company representative.

Negative Pulse System

The transmitter consists of a valve which, when opened, allows a small volume of mud to escape from the drill string into the annulus. The rapid opening and closing of this valve therefore creates a drop in standpipe

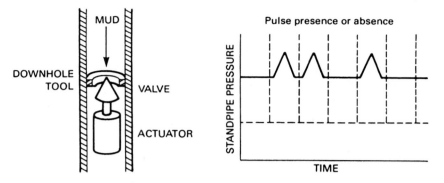

Fig. 8.6. Positive pulse system.

pressure that can be detected by the pressure transducer (Fig. 8.7). As with the positive system, a number of reference pulses precede the data pulses to set up the decoding process. The method used to decode the message varies between different companies. The presence or absence of a pulse within a time-frame, or the time interval between successive pulses, are two of the characteristics currently being used to interpret the negative pulse sequence. As with the positive pulse system, a manual interpretation of the sequence is possible by means of a chart recorder.

Continuous Wave Systems

Unlike the two previous methods, no distinct pulses are generated in this system. The transmitter is a rotary valve which consists of a pair of slotted discs mounted at right angles to the mud flow. One of the discs is stationary while the other is driven by a motor (Fig. 8.8). The constant speed of the motor creates a regular and continuous variation in pressure that is

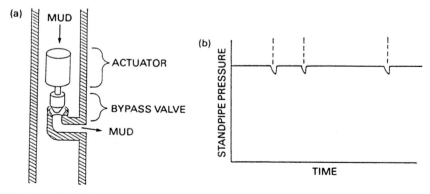

Fig. 8.7. Negative pulse system. (a) Presence or absence of pulse. (b) Time between pulses.

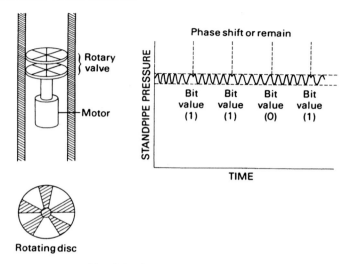

Fig. 8.8. Continuous wave system.

essentially a standing wave. This wave is used as a carrier to transmit the data to surface. When information is to be transmitted the speed of the motor is reduced so that the phase of the carrier wave is altered (i.e. reversed). The carrier wave is therefore modulated to represent the data required. The surface equipment detects these phase shifts in the pressure signal and translates this into a binary code. This is a more sophisticated telemetry system and offers a higher data rate than the previous two mud pulse methods. However, the complexity of both the downhole and surface components has limited its wider use.

POWER SOURCES

Since mud pulse telemetry systems have no wireline conductors running back to surface, the power source to operate the tool must be located downhole. Two forms of power source are being employed.

Batteries
These offer the advantage of being compact and reliable since they contain no moving parts. However, they do have a finite operational life and are temperature-dependent. They have been successfully used for applications in which only directional data are required. Their limited power output does not meet the requirements of multisensor tools (Fig. 8.9).

Turbine-alternator
With the trend towards multisensor tools, turbines are becoming more widely used to provide power for the MWD tool. The flow of mud through

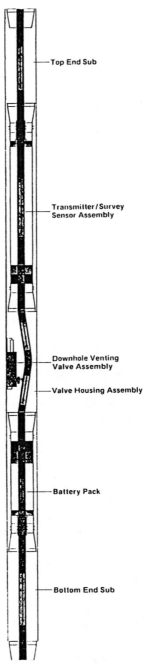

Fig. 8.9. MWD tool using a battery pack (courtesy of Eastman-Christensen).

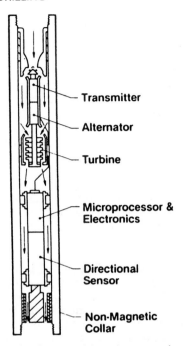

Fig. 8.10. MWD tool using a turbine alternator (courtesy of Teleco).

the tool is harnessed by the turbine blades, which rotate a shaft connected to an alternator (Fig. 8.10). The electrical power generated must be controlled by a voltage regulator. Although this system provides more power and longer operating life than a battery pack, power failures can occur if the turbine is damaged. To prevent damage a screen can be installed upstream of the turbine to filter out any debris in the mud. The screen may be positioned at the top of the drill string for ease of access if it requires to be emptied or removed to allow passage of wireline tools.

MWD SENSORS

All sensors used in the downhole tool must be rugged enough to withstand the harsh environment. Only one of the systems currently available has the facility to change-out damaged components (sensor package/electronics) without tripping the whole bottom hole assembly out of the hole. The failure of one sensor may not necessarily mean that the tool has to be pulled out. If the operator is satisfied that the other sensors are still providing the required information, the loss of one piece of data may not be significant (e.g. a gamma-ray failure in a directional well, when the operator places more importance on the directional data).

Directional Sensors

The directional sensors currently being used in MWD tools are triaxial magnetometers and accelerometers similar to those used in steering tools (Fig. 8.11). These sensors measure the required angles of inclination, azimuth and toolface. Since the magnetometers measure azimuth relative to Magnetic North, the correct magnetic declination must be applied to the results. The C axis is aligned with the axis of the tool, and the B axis defines the reference for measuring toolface angle. The angular offset between the B axis and the scribe line of the bent sub must be measured before running in the hole.

Both magnetometers and accelerometers give voltage outputs that have to be corrected by applying calibration coefficients. The corrected voltages can then be used to calculate the required directional angles (see Chapter 7, p. 120–1). As a check on the data being sent, the vector sum of the magnetic and gravitational components can be calculated and printed out with the survey results.

The sensors are powered up by the control system after some kind of signal has been sent from surface. The actual signal varies, depending on the particular system being used, but it is generally a change in mud pressure or a halt in drill string rotation. A transducer or motion sensor within the downhole tool recognizes this signal and initiates the survey. During the time when the sensors are actually taking the measurements the drill string must remain stationary for accurate results to be obtained. This period is generally less than 2 min., after which normal drilling can resume. The data are transmitted to surface while drilling ahead. The measurements of inclination azimuth and toolface are sent in a predetermined order. It generally takes 2–4 min. for transmission of a complete directional survey. The accuracy of the results is usually given as ±0.25° for inclination, ±2° for azimuth and ±3° for toolface although there is some slight variation in the figures quoted by MWD companies.

Gamma Ray Sensors

Gamma rays are emitted by radioactive elements such as isotopes of potassium, thorium and uranium. These elements are found more com-

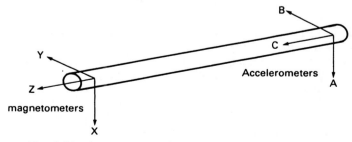

Fig. 8.11. Arrangement of sensors in directional package.

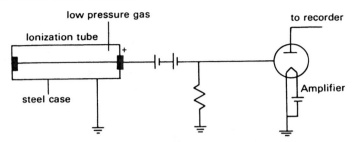

Fig. 8.12. Geiger–Muller tube.

monly in shales than in other rocks. By measuring the gamma-ray emission from a sequence of rocks it is therefore possible to identify shale zones. A gamma-ray sensor mounted in an MWD tool can detect this radiation as the bit drills through the formation. To be most effective in detecting changes of lithology, the gamma ray sensor should be positioned as close to the bit as possible, so that only a few feet of a new formation are drilled before the tool responds. For practical reasons, the distance between the bit and the gamma-ray sensor is about 6 ft. It should be remembered that only a small percentage of the gamma-rays being emitted will actually be detected, owing to attenuation in the mud and the drill collar.

There are two types of sensor used:

Geiger–Muller tube. This type of sensor consists of a cylinder that contains an inert gas at a fairly low pressure. A high-voltage electrode (±1000 V) runs through the centre of the chamber. As gamma-rays enter the chamber they cause ionization of the gas, creating a flow of fast-moving electrons towards the central electrode (Figure 8.12). The current of electrons can therefore be used to measure the amount of gamma-rays emitted from the formation.

Scintillation counter. The natural gamma-rays emitted by the formation pass through a sodium iodide crystal. The radiation excites the crystal, which produces a flash of light. The light emitted by the crystal strikes the photocathode and releases electrons. The electrons travel through a series of anodes, causing the emission of more electrons. This generates a voltage pulse which is proportional to the original flash of light (Fig. 8.13). The amount of radiation entering the sensor can therefore be measured by counting the number of pulses over a given time period. This is basically the same principle as is used in the gamma-ray sensor of a wireline logging tool.

The Geiger–Muller tube is not as accurate as the scintillation counter, since it can only detect a much smaller percentage of the total rays emitted. It does have the advantage, however, of being more rugged and reliable and being cheaper than the scintillation counter.

The measurements taken by the sensors are converted into a numerical code by the MWD control system and stored until ready to be transmitted. As with directional measurements the data is sent as a series of mud pulses

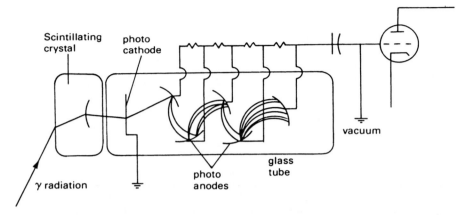

Fig. 8.13. Scintillation counter.

to the surface. The end result is the gamma-ray response (measured in counts per second) plotted versus depth in the form of a continuous log.

Resistivity Sensor

Resistivity is a measure of the formation's resistance to the flow of electric current. The response from the formation will depend on the fluid content of the pore space (oil and gas act as insulators, while brine is a conductor). The resistivity sensor on an MWD tool has been adapted from the equivalent wireline logging tool (16-in. normal device). Two electrodes are mounted on an insulating rubber sleeve on the outside of the MWD tool (Fig. 8.14). The electric current emitted by the upper electrode passes through the formation and is detected by the lower electrode. The actual

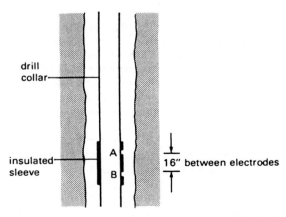

Fig. 8.14. Resistivity sensor (16-in. short normal device).

response is also affected by borehole dimensions, mud invasion and bed thickness. Certain correction factors must be applied to compensate for these effects. This type of sensor will not be effective in boreholes in which oil-based mud is being used. An induction type sensor has been developed for oil-based muds and can be incorporated in an MWD tool. As with the gamma-ray sensor, the resistivity device should be installed close to the bit to give a fast response to formation changes.

Temperature Sensor

The temperature sensor is usually mounted on the outside wall of the drill collar, and therefore monitors the annulus mud temperature. The sensing element may be a strip of metal (e.g. platinum) whose electrical resistance changes with temperature. The sensor can be calibrated to measure temperatures ranging from 50 to 350°F.

Downhole WOB/Torque

These measurements are made by a system of sensitive strain gauges mounted on a special sub placed close to the bit. The strain gauges will detect axial forces for WOB and torsional forces for torque. By placing pairs of gauges on opposite sides of the sub, any stresses due to bending can be eliminated. The sub must also be designed to compensate for the effects of temperature and pressure.

Turbine RPM

When drilling with a downhole turbine, the actual speed at which the bit is turning is not known at surface. The only effective way of monitoring the rpm is to use a turbine tachometer linked to an MWD system to provide real time data. The downhole sensor consists of a 2-in. diameter probe that is placed very close to the top of the rotating turbine shaft. On top of the shaft are mounted two magnets 180° apart. As the shaft rotates, an electric coil within the probe picks up voltage pulses due to the magnets (Fig. 8.15). By counting the number of pulses over a certain interval, the turbine speed in rpm can be calculated. This information is encoded as a series of mud pulses that are transmitted at intervals to surface to let the driller know how the rpm is changing.

SURFACE SYSTEM

Most MWD companies use a fairly similar set of surface equipment to interpret, record and display the data measured by the downhole sensors. The amount of surface equipment will depend on the company and the number of different parameters being measured. If only directional data are required, the MWD company may only require a standpipe pressure transducer and a receiver/processing system that can be installed in the driller's dog-house. If formation evaluation data and drilling parameters

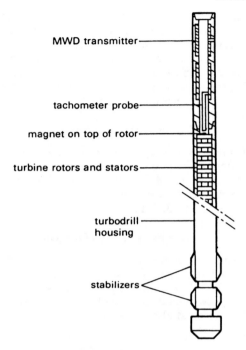

MWD transmitter

tachometer probe

magnet on top of rotor

turbine rotors and stators

turbodrill
housing

stabilizers

Fig. 8.15. Turbine rpm sensor.

are also required, it is more practical to house all the electronic equipment
and plotting devices in the mud-logging cabin or purpose-built unit located
on the pipe deck. The basic components of the surface system are shown in
Fig. 8.16. These are described as follows.

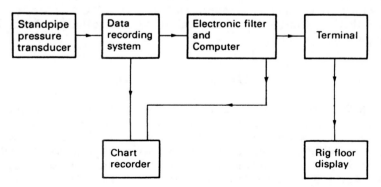

Fig. 8.16. Main components of surface system.

Standpipe Pressure Transducer

On most drilling rigs the standpipe manifold has a number of pressure taps where gauges may be installed. The transducer can be installed at a convenient point by removing one of these gauges. Inside the transducer is a sensitive diaphragm that detects variation in pressure and converts these hydraulic pulses to electrical voltage pulses. The voltage output is relayed to the rest of the surface equipment by means of an electric cable.

Electronic Filters—Amplifiers—Surface Computer

As well as detecting the MWD pulses, the transducer will also respond to pressure variations caused by the rig pumps or downhole motors. This background noise makes it difficult to identify the MWD pulses. It is sometimes possible to alter the speed of the rig pumps to reduce this interference. Pulsation dampeners on the rig pumps should prevent any large variations in discharge pressure.

The signal from the transducer can be improved by filtering out these unwanted pressure fluctuations. With the expected frequency of the MWD signal known, the electronic filters can be set to eliminate any other frequencies above or below this pre-set range. The signal can then be further enhanced by amplifying it to show the distinctive peaks or troughs. The enhanced sequence of pulses is then fed into the surface computer which is programmed to recognize the reference pulses and then decode the data pulses. The final results are listed on a print-out at the computer terminal and recorded on magnetic tape.

Directional results (azimuth, toolface and inclination) are relayed to the rig floor where they are displayed on a panel for the convenience of the directional driller. Formation evaluation data are printed out continuously as drilling proceeds. A depth-tracking system must also be used to plot gamma-ray or resistivity response against depth.

Directional data, formation data and downhole drilling parameters are transmitted in a pre-determined sequence. The sequence and frequency of the measurements varies with different manufacturers and also depends on how the tool is being operated. For example during a steering run only survey data may be transmitted; during rotary drilling gamma ray and resistivity may be transmitted.

DRILLING WITH AN MWD SYSTEM

Pre-job Requirements

An operating company wishing to use an MWD system must first decide:

(a) what the basic data requirements are;
(b) which systems currently available can meet these requirements;
(c) whether the limitations and specifications of the system under consideration are acceptable for this application (in terms of accuracy, data rate, temperature, etc.).

Where several different tools meet all the job requirements, some operators have a deliberate policy of trying out all the different systems to assess reliability and cost-effectiveness. The MWD company chosen to provide the service will require certain information from the operator:

(a) when does the operator anticipate using the MWD tool (sufficient time must be allowed for workshop testing, transportation of equipment and personnel, rigging-up time at the well site);
(b) the expected flow rates, pump pressures, bit nozzle sizes, mud weight and other BHA components (this will affect the choice of internal MWD components fitted to the tool).

All the components should be thoroughly checked before the tool leaves the workshop. Directional sensors should be tested on a calibration stand that can simulate the expected azimuth and inclination. Some companies are able to install all the necessary downhole components inside the non-magnetic drill collar before leaving the workshop. It may be necessary to supply more than one tool so that, if there is a failure, another tool is already on the rig to use as a back-up.

Rigging-up and Surface Checks

All MWD systems have been designed to allow a fairly simple rig-up so that normal drilling operations are not seriously held up. The service company will supply all the necessary downhole and surface equipment. To rig-up and run an MWD system usually requires two service company engineers.

The surface system requires the installation of the pressure transducer at a convenient point in the standpipe manifold. Cables are then laid to the MWD unit situated in the dog-house or on the pipe deck. The rigging-up of the surface systems may require the assistance of the rig electrician or safety officer. A rig-floor display is connected up to allow the directional driller to monitor changes in the measured angles.

If the downhole components are already mounted in the drill collar, the rig-up is a simple operation, provided the collar is handled carefully. If the downhole tool has to be assembled at the rig, each individual component can be checked out. As the collar is being made up, careful attention should be paid to ensure the correct torque is applied. Rough handling of tongs and slips may damage the sensitive components inside the collar.

If the MWD tool is being used with a downhole motor and bent sub, the offset toolface must be measured. This is the angular difference between the actual toolface (as defined by the scribe line of the bent sub) and the toolface of the MWD tool (as defined by the position of the B axis). This may be done at the rig floor by a measuring tape or a purpose-built protractor (Fig. 8.17). The offset toolface must be stored in the surface computer so that the MWD toolface can be automatically converted to actual toolface. Likewise, the magnetic declination should also be stored in the computer so that magnetic azimuth can be converted to true azimuth.

While the tool is suspended at the rig floor it should be function-tested to

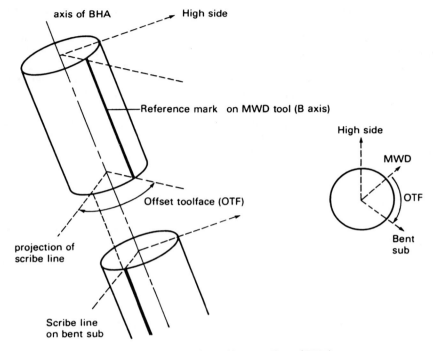

Fig. 8.17. Measuring offset toolface (OTF).

ensure that all components are operating correctly. A useful test on the reliability of the tool is to take a survey at a particular depth on every trip into the hole. This is known as a benchmark survey and all the results should show close agreement. The benchmark survey is usually taken below the casing shoe at a shallow depth, so that if the tool is not giving the expected result it can be tripped out without wasting too much rig time.

Normal Surveying Procedure

For directional MWD there are two modes of operation:

(a) Rotary drilling, in which static surveys (azimuth and inclination) are taken after each connection, or at closer intervals if required.
(b) Steering runs, in which a bent sub and downhole motor are being used. In this case it is more important to monitor the toolface as the bit is drilling.

In some tools the operator must specify which mode is required; in other tools both static and dynamic surveys are transmitted in sequence. The normal procedure for a static survey is to drill to kelly down, make the connection and take the survey with the bit about 5 ft off-bottom. During the 2 min. required for the sensors to measure the data, the pipe must remain stationary. While drilling, the pulses are transmitted to surface and the survey is displayed within about 4 min.

During steering runs, the updates of toolface are transmitted at short intervals (every minute). Normally the toolface is referenced to the High Side of the hole (i.e. gravity toolface). However at low inclinations (less than 8°) the toolface will be referenced to Magnetic North (i.e. magnetic toolface). Magnetic toolface, however, cannot be used near casing, since there will be magnetic interference. The same is true of azimuth. For any surveying or orientation of deflecting tools near casing, therefore, gyroscopic tools must be used. MWD tools can only be used where the operator is satisfied that there is no magnetic interference.

APPLICATIONS OF MWD

The applications of MWD can be divided into three broad categories which are described as follows.

Directional Surveying

This accounts for approximately 70% of all MWD jobs. In the North Sea, MWD has become the standard method of monitoring the wellpath as drilling proceeds. The major benefits of MWD in directional drilling are that:

(a) Valuable rig time is saved when taking surveys, owing to the elimination of conventional wireline techniques.
(b) Orientation of toolface for steering runs is made much easier, as is monitoring of toolface while drilling with no wireline problems.
(c) Less time is spent with the drill string in a stationary position, thereby reducing the risk of stuck pipe.
(d) Closer density of surveys is possible without a great loss of rig time; therefore there is better monitoring of the wellpath.
(e) The effects that changing drilling parameters or formation changes may have on the well path can be detected very much quicker, reducing the risk of severe dog-legs and the need to make correction runs.

The cost-effectiveness of an MWD system and its reliability are of prime importance to the operator. Several studies have shown that MWD can save considerable rig time when compared to conventional wireline surveying techniques. This potential benefit will of course be offset by any tool failures that force the operator to waste time by pulling the MWD tool. Over the past few years, the average failure rate—in terms of mean circulating time between failures—has improved from around 20 hours to 250 hours. This increase in reliability, and the cost reduction due to keener competition, have firmly established MWD as an integral part of directional drilling operations.

Formation Evaluation

Logging-while-drilling with gamma-ray and resistivity sensors is becoming more popular, but it is much more difficult to justify in strict economic terms. This is because the operator will probably wish to run a complete suite of wireline logs in any case. The incremental benefits of having logging data as the well is being drilled must be investigated. Some of these benefits are listed as follows:

(a) Selection of casing points using gamma ray log to identify shale zones.
(b) Picking the top of the reservoir to begin coring operations.
(c) Correlation with other neighbouring wells as drilling proceeds.
(d) Identification of troublesome zones.
(e) Ability to run logs in high-angled wells where wireline methods may not be suitable.
(f) At least some formation data is available if the hole is lost before wireline logs are run.
(g) Resistivity logs can detect the presence of shallow gas zones.
(h) Formation pressures can be evaluated while drilling using data from gamma-ray and resistivity logs.

An example of an MWD log is shown on Fig. 8.18.

The increasing credibility of MWD logs is such that in some areas MWD has replaced some intermediate wireline logs. One major drawback at the moment is the absence of an MWD porosity sensor. There has also been much debate over the comparison between MWD logs and wireline logs. While some very close agreement has been shown in certain examples, it should be remembered that there are some important differences in the logging techniques:

(a) The logging speeds are very different (MWD may be 10–100 ft/h, wireline may be 1800 ft/h). The resolution of the logs will therefore also be different.
(b) The condition of the hole may have changed owing to the effects of mud invasion.
(c) The centralization of the tools may be different.
(d) There may be a difference in the type of sensor used (e.g. Geiger–Muller tube in MWD as opposed to scintillation counter in wireline tool).
(e) Signal attenuation due to the drill collar will affect results.
(f) The MWD-GR log is measured in counts per second, while the wireline log is in API units.

Drilling Parameters

Of the three main applications of MWD, the use of sensors to measure downhole drilling parameters is perhaps the most difficult to justify in terms of cost-effectiveness. This is because these sensors are not replacing some other system which was more expensive to run, such as survey tools or logging tools.

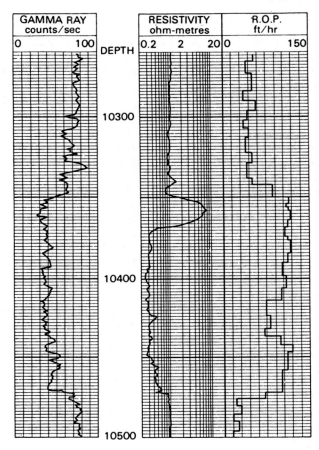

Fig. 8.18. MWD log showing gamma ray and resistivity response. Rate of penetration with depth is also given for correlation.

The major benefit of having downhole sensors for WOB, rpm or torque is that the measurements are actually being made where it matters—at the bit. Surface indicators cannot be assumed to be reliable in certain circumstances (e.g., in a highly deviated hole, downhole WOB may be as low as 20% of the value indicated on surface, owing to wall friction). This discrepancy between downhole and surface measurements may give some indication of hole problems (e.g., surface torque being much greater than bit torque suggests possible stuck-pipe problems as opposed to a locked cone at the bit).

Apart from providing useful indicators to the driller as the hole is being drilled, these sensors have a wider application in providing the input to optimization programs, rather than taking inaccurate surface data. In some instances (e.g. turbodrilling) an MWD sensor is the only available method of optimizing drilling operations. As yet the benefits of downhole drilling sensors remain largely undiscovered, but they will certainly have a part to play in the future.

QUESTIONS

8.1. List four methods of sending information from downhole to surface, outlining the main advantages and disadvantages of each.

8.2. In mud pulse telemetry, how does the positive pulse method differ from the negative pulse method?

8.3. The following data were obtained from the output of an MWD survey:

Accelerometer voltages
$G_x = -0.0132$
$G_y = 0.0157$
$G_z = 1.0141$

Magnetometer voltages
$H_x = 0.1062$
$H_y = 0.2510$
$H_z = 0.9206$

The offset toolface = 0° and the magnetic declination = 7° W.
From this data calculate:
(a) inclination;
(b) azimuth;
(c) gravity toolface.
Comment on the reliability of the gravity toolface in this case.

8.4. After making up the BHA for a steering run, the offset toolface must be determined before running in the hole. For an 8-in. OD tool the service company engineer measures the length from the reference axis of the MWD tool to the scribe line of the bent sub. If this distance is 9.5-in. what is the offset toolface?

8.5. List the applications of gamma-ray and resistivity sensors in an MWD tool that may be useful in drilling operations.

8.6. Discuss the relative merits of the two types of gamma-ray sensors that are presently used in MWD tools.

8.7. What factors should be considered when comparing MWD gamma-ray logs with wireline gamma-ray logs?

FURTHER READING

"Second generation MWD tool passes field tests", *Oil and Gas Journal*, Report, 21 February 1983.

"Mud pulse MWD systems report", M. Gearthart, K. A. Ziemer and O. M. Knight, S.P.E. paper no. 10053.

"Downhole telemetry from the user's point of view", A. W. Kamp, *Journal of Petroleum Technology*, October 1983.

"MWD logs provide fast, accurate data", D. F. Coope, and W. E. Hendricks, *World Oil*, 1 August 1985.

"MWD tools improve drilling performance", S. D. Moore, *P.E.I.*, February 1986.

"Drilling interpretation from MWD data", W. G. Lesso and I. G. Falconer, Conference on MWD, London, 6 June 1986.

"High data rate drilling telemetry system", E. B. Denison, S.P.E. paper no. 6775.

"Applications of MWD", D. R. Tanguy and W. A. Zoeller, S.P.E. paper no. 10324.

"Telemetry—MWD—The second tier benefits", R. Newton, R. L. Kite and F. A. Stone, S.P.E. paper no. 9224.

"MWD tool performance in the North Sea", D.J.W. Knox. S.P.E. paper no. 16523.

Chapter 9

SURVEY CALCULATIONS

The results of a directional survey are given in terms of azimuth and inclination of a borehole at a certain depth. In this chapter it will be assumed that all the necessary corrections (magnetic declination, gyro drift) have been applied to the survey results. This information must then be analysed to calculate the actual position of the wellbore at that survey station relative to the surface location. In order to do this the incremental distances ΔV, ΔE, ΔN between the successive survey stations must be calculated. With the coordinates of the upper station known, the coordinates of the lower station can be found by addition. The horizontal coordinates of a point are referred to as the "Northing" (or latitude) and the "Easting" (or departure).

The inclination and azimuth at each survey station define two vectors that are tangential to the wellbore trajectory. The inclination vector lies in the vertical plane, while the azimuth lies in the horizontal plane. The only other piece of information available is the course length (the difference in survey depths) between the two stations. It is necessary therefore to assume some kind of idealized wellpath between the upper and lower stations. Various different kinds of geometrical models have been used, with each model generating a number of mathematical equations. The assumed wellpath may simply be a straight line joining the two survey stations or it could be some kind of curved line defined by the end points.

The accuracy of the final coordinates will naturally depend on how well the assumed trajectory used in the model approximates to the actual trajectory in the borehole. The position of the wellbore must be known precisely at critical stages during drilling (e.g. when kicking off near existing wells). An operating company will usually adopt one method of calculating the wellbore position and apply this model to all surveys throughout the length of the well. In order to be consistent it is important that same model be applied to all the other wells drilled from that platform.

CALCULATION TECHNIQUES

The more common methods that have been adopted are described in the following paragraphs. In this section the following symbols will be used. Inclination and azimuth will be represented by α and β respectively, with the subscript 1 denoting the upper station and 2 denoting the lower station. The course length L between the two stations is equal to the difference in survey depths. The symbols ΔV, ΔN and ΔE are the incremental distances between stations along the three axes (i.e. vertical, northing and easting). In each method the course length L has to be resolved in both the vertical and horizontal planes.

Tangential Method

In this model the wellpath is assumed to be a straight line defined by the inclination and azimuth at the lower survey station (Fig. 9.1). Notice that the angles measured at the upper station are not used in the analysis. From Fig. 9.1 it can be seen that:

$$\Delta V = L \cos \alpha_2$$
$$\Delta N = L \sin \alpha_2 \cos \beta_2$$
$$\Delta E = L \sin \alpha_2 \sin \beta_2$$

This method clearly gives large errors in wellbore position when the trajectory is changing significantly between stations. In a directional well, where even over relatively short intervals there can be significant changes in azimuth and inclination, this method of calculation is not recommended.

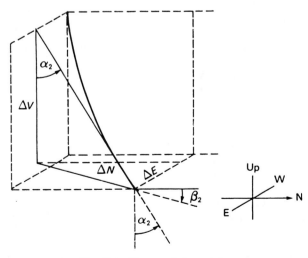

Fig. 9.1. Tangential model.

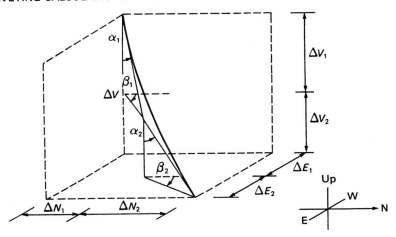

Fig. 9.2. Balanced tangential model.

Balanced Tangential Method

This method assumes that the actual wellpath can be approximated by two straight line segments of equal length. The upper segment is defined by α_1 and β_1, while the lower segment is defined by α_2 and β_2. The length of each segment $= L/2$. From Fig. 9.2. it can be seen that:

$$\Delta V = \tfrac{1}{2}L \cos \alpha_1 + \tfrac{1}{2}L \cos \alpha_2$$
$$= \tfrac{1}{2}L(\cos \alpha_1 + \cos \alpha_2)$$
$$\Delta N = \tfrac{1}{2}L \sin \alpha_1 \cos \beta_1 + \tfrac{1}{2}L \sin \alpha_2 \cos \beta_2$$
$$= \tfrac{1}{2}L(\sin \alpha_1 \cos \beta_1 + \sin \alpha_2 \cos \beta_2)$$
$$\Delta E = \tfrac{1}{2}L \sin \alpha_1 \sin \beta_1 + \tfrac{1}{2}L \sin \alpha_2 \sin \beta_2$$
$$= \tfrac{1}{2}L(\sin \alpha_1 \sin \beta_1 + \sin \alpha_2 \sin \beta_2)$$

This method is considerably more accurate than the tangential method, since it does take into account both sets of survey data. It can be further improved by applying a ratio factor (see Minimum Curvature Method).

Average Angle Method

This method assumes only one straight line that intersects both upper and lower stations. The straight line is defined by averaging the inclination and azimuth at both stations. From Fig. 9.3:

$$\Delta V = L \cos \left(\frac{\alpha_1 + \alpha_2}{2}\right)$$

$$\Delta N = L \sin \left(\frac{\alpha_1 + \alpha_2}{2}\right) \cos \left(\frac{\beta_1 + \beta_2}{2}\right)$$

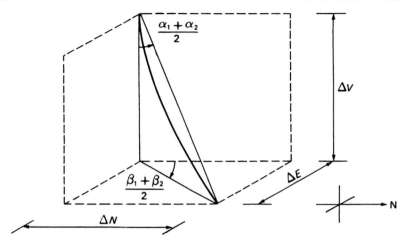

Fig. 9.3. Average angle model.

$$\Delta E = L \sin\left(\frac{\alpha_1 + \alpha_2}{2}\right)\sin\left(\frac{\beta_1 + \beta_2}{2}\right)$$

This is a very popular method, since it yields accurate results and is fairly simple to use with the aid of a hand-held calculator. For this reason it is often used at the well-site, provided the survey stations are not far apart. This method has been shown to be unreliable in near vertical wells.

Radius of Curvature Method

This method assumes that the wellpath is not a straight line but a circular are when viewed in both the vertical and horizontal planes. The arc is tangential to the inclination and azimuth at each survey station (Fig. 9.4). The wellpath can therefore be described as an arc in the vertical plane, which is wrapped around a right vertical cylinder.

In the vertical plane
$\stackrel{\frown}{\text{AOB}} = \alpha_2 - \alpha_1$, therefore

$$\frac{\alpha_2 - \alpha_1}{360} = \frac{L}{2\pi R_v}$$

the radius in the vertical plane, R_v, can be found from

$$R_v = \frac{L}{\alpha_2 - \alpha_1}\left(\frac{180}{\pi}\right)$$

$$\Delta V = R_v \sin \alpha_2 - R_v \sin \alpha_1 = R_v(\sin \alpha_2 - \sin \alpha_1)$$

Substituting for R_v, the vertical increment becomes

$$\Delta V = \frac{L}{\alpha_2 - \alpha_1}\left(\frac{180}{\pi}\right)(\sin \alpha_2 - \sin \alpha_1)$$

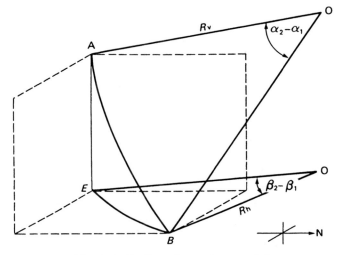

Fig. 9.4. Radius of curvature model.

The horizontal increment (ΔH) can be found from

$$\Delta H = R_v(\cos \alpha_1 - \cos \alpha_2)$$

In the horizontal plane

$\widehat{EOB} = \beta_2 - \beta_1$, therefore

$$\frac{\beta_2 - \beta_1}{360} = \frac{\Delta H}{2\pi R_h}$$

The radius in the horizontal plane, R_h, becomes

$$R_h = \frac{\Delta H}{\beta_2 - \beta_1} \left(\frac{180}{\pi}\right)$$

$$\Delta N = R_h \sin \beta_2 - R_h \sin \beta_1 = R_h(\sin \beta_2 - \sin \beta_1)$$

Substituting for R_h:

$$\Delta N = \frac{\Delta H}{\beta_2 - \beta_1} \left(\frac{180}{\pi}\right)(\sin \beta_2 - \sin \beta_1)$$

Substituting for ΔH:

$$\Delta N = \frac{R_v(\cos \alpha_1 - \cos \alpha_2)}{\beta_2 - \beta_1} \left(\frac{180}{\pi}\right)(\sin \beta_2 - \sin \beta_1)$$

Substituting for R_v:

$$\Delta N = \frac{L}{\alpha_2 - \alpha_1} \left(\frac{180}{\pi}\right)^2 \frac{(\cos \alpha_1 - \cos \alpha_2)(\sin \beta_2 - \sin \beta_1)}{\beta_2 - \beta_1}$$

Similarly for ΔE:

$$\Delta E = \frac{L}{\alpha_2 - \alpha_1}\left(\frac{180}{\pi}\right)^2 \frac{(\cos \alpha_1 - \cos \alpha_2)(\cos \beta_1 - \cos \beta_2)}{\beta_2 - \beta_1}$$

This method does provide better results than average angle in sections of the hole where the path is closer to a circular arc (e.g. during kick-off). However, it assumes a constant radius, which may not be true over longer intervals. In straight sections of hole there are computational problems due to division by zero.

Minimum Curvature Method

This method is really an extension of the balanced tangential method. Rather than assuming that the actual wellpath is approximated by two straight line segments, this method replaces the straight lines by a circular arc. This is done by applying a ratio factor based on the amount of bending in the wellpath between the two stations (dog-leg angle). The dog-leg angle can be calculated from

$$\phi = \cos^{-1}[\cos \alpha_1 \cos \alpha_2 + \sin \alpha_1 \sin \alpha_2 \cos(\beta_2 - \beta_1)]$$

The derivation of this formula is given in the next section. From Fig. 9.5 it can be seen that the ratio factor F can be calculated from

$$F = \frac{AB + BC}{\text{arc AC}}$$

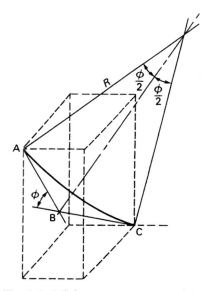

Fig. 9.5. Minimum curvature model.

and

$$AB = BC = R \tan(\phi/2)$$

and

$$\frac{AC}{2\pi R} = \frac{\phi}{360} \Leftrightarrow AC = \frac{\pi R \phi}{180}$$

Therefore,

$$F = \frac{2}{\phi} \left(\frac{180}{\pi} \right) \tan \left(\frac{\phi}{2} \right)$$

This ratio factor is then applied to the results of ΔV, ΔN and ΔE as given by the balanced tangential method (p. 157). The equations for the minimum curvature method can be summarized as follows:

$$\Delta V = F \frac{L}{2} (\cos \alpha_1 + \cos \alpha_2)$$

$$\Delta N = F \frac{L}{2} (\sin \alpha_1 \cos \beta_1 + \sin \alpha_2 \cos \beta_2)$$

$$\Delta E = F \frac{L}{2} (\sin \alpha_1 \sin \beta_1 + \sin \alpha_2 \sin \beta_2)$$

The minimum curvature method is the one most often adopted for directional surveying calculations. Owing to the more complicated mathematics involved, this method is more suited to computer techniques. Various programs have been written for hand-held programmable calculators, so that this method can be used at the well site.

From Table 9.1 it can be seen that for a typical set of survey data the different methods give very close agreement, with the exception of the tangential method.

DOG-LEG ANGLE

To derive the formula to calculate the dog-leg angle used in the previous section, consider the survey stations shown in Fig. 9.6. At the upper station the inclination and azimuth have been measured as α_1 and β_1. At the lower station the corresponding angles are α_2 and β_2. These angles define the two straight line segments whose lengths are L_1 and L_2. The change in total angle (ϕ) between these two segments is shown as in the diagram. The size of the angle ϕ can be determined by considering the triangle bounded by the lines L_1, L_2 and L_3. The true length of L_3 can be determined by considering the true vertical depth and the horizontal displacement between stations 1 and 2:

$$\Delta V = L_1 \cos \alpha_1 + L_2 \cos \alpha_2$$

ΔH can be derived from the horizontal projection of L_1 and L_2 by applying

TABLE 9.1   Summary of Results Using Each of the Methods

Notice that each calculation method gives fairly close agreement, with the exception of the tangential method which underestimates ΔV, and overestimates ΔN and ΔE.

	M.D.	INC	AZI	ΔV	ΔN	ΔE
Tangential	2000	2.0	045			
	2090	4.5	050	89.72	4.54	5.41
	2180	7.5	053	89.23	7.07	9.38
	2270	10.5	048	88.49	10.98	12.19
	2360	14.0	055	87.33	12.49	17.84
Balanced	2000	2.0	045			
tangential	2090	4.5	050	89.83	3.38	3.82
	2180	7.5	053	89.48	5.80	7.40
	2270	10.5	048	88.86	9.02	10.79
	2360	14.0	055	87.91	11.73	15.01
Average	2000	2.0	045			
angle	2090	4.5	050	89.86	3.45	3.76
	2180	7.5	053	89.51	5.86	7.36
	2270	10.5	048	88.89	8.96	10.86
	2360	14.0	055	87.95	11.89	14.94
Radius of	2000	2.0	045			
curvature	2090	4.5	050	89.85	3.45	3.76
	2180	7.5	053	89.50	5.86	7.36
	2270	10.5	048	88.88	8.95	10.86
	2360	14.0	055	87.94	11.88	14.93
Minimum	2000	2.0	045			
curvature	2090	4.5	050	89.84	3.38	3.82
	2180	7.5	053	89.50	5.80	7.40
	2270	10.5	048	88.88	9.02	10.79
	2360	14.0	055	87.94	11.73	15.02

the cosine rule:

$$(\Delta H)^2 = (L_1 \sin \alpha_1)^2 + (L_2 \sin \alpha_2)^2 - 2L_1 \sin \alpha_1 \, L_2 \sin \alpha_2 \cos(180 - \Delta)$$

where Δ = change in azimuth = $\beta_2 - \beta_1$ (in the horizontal plane).

Since $\cos(180 - \Delta) = -\cos \Delta$, this can be re-written as:

$$(\Delta H)^2 = (L_1 \sin \alpha_1)^2 + (L_2 \sin \alpha_2)^2 + 2L_1 \sin \alpha_1 \, L_2 \sin \alpha_2 \cos \Delta$$

Length L_3 can therefore be determined from

$$(L_3)^2 = (\Delta V)^2 + (\Delta H)^2$$

Substituting for ΔV and ΔH:

$$(L_3)^2 = (L_1)^2 + (L_2)^2 + 2L_1L_2(\cos \alpha_1 \cos \alpha_2 + \sin \alpha_1 \sin \alpha_2 \cos \Delta)$$

Applying the cosine rule to the triangle bounded by L_1, L_2 and L_3:

$$\cos(180 - \phi) = \frac{(L_1)^2 + (L_2)^2 - (L_3)^2}{2L_1L_2}$$

$$= -(\cos \alpha_1 \cos \alpha_2 + \sin \alpha_1 \sin \alpha_2 \cos \Delta)$$

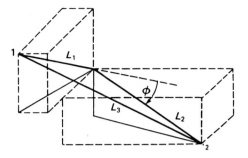

Fig. 9.6. Diagram to illustrate dog-leg severity.

or

$$\cos \phi = \cos \alpha_1 \cos \alpha_2 + \sin \alpha_1 \sin \alpha_2 \cos \Delta$$

In its most common form, the dog-leg angle can therefore be expressed as

$$\phi = \cos^{-1}[\cos \alpha_1 \cos \alpha_2 + \sin \alpha_1 \sin \alpha_2 \cos(\beta_2 - \beta_1)]$$

The dog-leg severity is calculated by dividing the dog-leg angle by the course length, and expressing this in terms of degrees per 100 ft:

$$\text{DLS} = 100 \frac{\phi}{L}$$

where ϕ = dog-leg angle (degrees)
$\quad\ \ L$ = difference in survey depth (ft)
$\quad$ DLS = dog-leg severity (degrees/100 ft).

EXAMPLE 9.1

Calculate the dog-leg severity between these two survey stations:

MD	α	β
2000	4.5°	148°
2031	5.5°	145°

$$\cos \phi = \cos 4.5° \cos 5.5° + \sin 4.5° \sin 5.5° \cos(145° - 148°)$$
$$\Leftrightarrow \phi = 1.03°$$

$$\text{DLS} = 100 \frac{1.03}{31} = 3.3° \text{ per 100 ft}$$

VERTICAL SECTION

Using one of the methods already outlined, the coordinates of the various survey stations can be plotted on the horizontal plane as shown in Fig. 9.7.

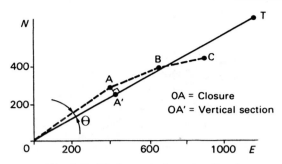

Fig. 9.7. Horizontal plan of wellpath.

To obtain the vertical view of the well profile, one particular plane has to be selected so that all the points can be plotted on a common axis corresponding to a particular direction. The plane usually selected is the one that contains both the target and the reference origin. All the points shown in Fig. 9.7 must therefore be projected onto this line. The length OA' in Fig. 9.7 is known as the vertical section and can be calculated as follows:

$$\text{length OA} = (N_a^2 + E_a^2)^{1/2}$$

where OA = closure of point A
N_a = northing of A
E_a = easting of A.

$$\text{angle A}\hat{\text{O}}\text{N} = \tan^{-1}(E_a/N_a)$$
$$\text{angle T}\hat{\text{O}}\text{N} = \tan^{-1}(E_t/N_t)$$

where T$\hat{\text{O}}$N = target bearing
N_t = northing of the target
E_t = easting of the target.

Let $\theta = $ T$\hat{\text{O}}$N $-$ A$\hat{\text{O}}$N:

$$\text{vertical section} = \text{OA}' = \text{OA} \cos \theta$$

With the TVD and the vertical section for each point known, the profile on the vertical plane can be plotted.

EXAMPLE 9.2

In Table 9.2 the coordinates of three successive survey stations are given. These correspond to the points shown in Fig. 9.7. If the target bearing is 060°, calculate the vertical section for each point and plot the vertical profile of the well.

For point A:

$$\theta = 60° - \tan^{-1}(400/300) = 60° - 53.1° = 6.7°$$
$$\text{OA}' = (300^2 + 400^2)^{1/2} \cos 6.7° = 496.59 \text{ ft}$$

TABLE 9.2   Coordinates of Survey Stations
(Target bearing = 060° (see Fig. 9.7)

Point	Northing	Easting	True vertical depth
A	300	400	1600
B	400	650	1800
C	450	900	2000

For point B:

$$\theta = 60° - \tan^{-1}(650/400) = 60° - 58.4° = 1.6°$$

$$OB' = (400^2 + 650^2)^{1/2} \cos 1.6° = 762.92 \text{ ft}$$

For point C:

$$\theta = 60 - \tan^{-1}(900/450) = 60° - 63.4° = -3.4°$$

$$OC' = (450^2 + 900^2)^{1/2} \cos(-3.4) = 1004.46 \text{ ft}$$

These points can now be plotted against TVD as shown in Fig. 9.8.

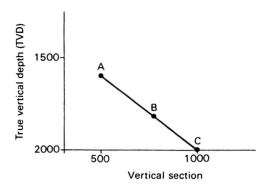

Fig. 9.8. View of the wellpath in vertical plane.

ERRORS IN SURVEYING

There are several sources of error that can lead to inaccuracy in determining the position of the borehole. Despite the use of sophisticated surveying instruments, the coordinates of the borehole can never be determined exactly. It is important, however, that the errors involved are quantified in such a way that it is possible to specify the borehole position within certain tolerance limits. From knowledge of the accuracy to which the depth, inclination and azimuth at the survey station can be measured, an area of uncertainty around the survey station can be defined. This takes the form of an ellipsoid, within which the actual borehole position lies.

The degree of uncertainty that can be tolerated depends on the particular application. To successfully locate the target in a conventional directional well it may be permissible to have a lateral error of 10 ft per 1000 ft drilled. When kicking off from a multiwell platform, however, it may be necessary to limit the error to 2 ft per 1000 ft drilled to avoid the risk of collision with adjacent wells. In the case of a relief well, the bottom hole location must be within 50 ft or less of the target. To achieve high levels of accuracy, the surveying programme must be carefully designed, in terms both of the selection of instruments and of the way in which they are run.

Sources of Survey Errors

Choice of algorithm used to calculate position

As described earlier, there are a number of different methods of calculation that can be employed. In a directional well, in which azimuth and inclination are likely to change at various stages in the wellpath, the tangential method is clearly not suitable. Provided the distance between surveys is not large, there is little difference in the results obtained from the remaining methods (see Table 9.1). Most operators have adopted the minimum curvature method as being the most accurate.

Errors related to the instrument itself

In both magnetic and gyroscopic devices the operating mechanism that measures the required angles has inherent inaccuracies. The magnetic compass will respond to any magnetic field present. In some cases a local magnetic field due to the steel drill string can cause a 10° error in the compass reading. A conventional gyroscope relies on accurate alignment on surface with the reference direction. Any error in setting up the bench-mark on the rig to define the reference direction, or in aligning the spin axis of the gyro to that direction, will affect all the subsequent surveys taken with that instrument. The gimbal bearings within the instrument are not perfectly balanced and may also lead to errors.

Errors related to the borehole environment

If the downhole pressure and temperature exceed the specifications of the survey tool being used, then the mechanism may not be able to provide reliable results. The specifications may also impose limits on inclination (e.g. gyros cannot normally be run in wells whose inclination exceeds 70°). At high angles of lattitude the reduced horizontal component of the Earth's magnetic field will affect the reliability of a magnetic compass. Drilling in an east–west direction will also reduce the reliability of a magnetic compass.

Misalignment of the survey tool with the axis of the borehole is another source of error. This is caused by poor centralization of the tool within the drill collar, or the bending of the drill collars themselves within the borehole.

Errors related to the survey depth

As well as reliable values for azimuth and inclination, the survey depth must also be accurately measured. Errors can arise from an incorrect tally of drill pipe lengths, or inaccurate wireline measurements. It is also possible for the baffle plate to have been installed at the wrong position within the BHA, leading to a discrepancy between the actual survey depth, and the assumed survey depth.

Errors in reading or reporting survey results
Single shot and multishot pictures can be difficult to read and mistakes can easily be made, especially at low inclinations. The correction for magnetic declination is also sometimes applied incorrectly or omitted completely in some cases.

Recommendations to Reduce Survey Errors

In order to minimize the errors involved, all surveying instruments must be checked and calibrated prior to running in the hole. The choice of instrument is dictated by the accuracy requirements and the operating limits specified by the manufacturer. The checks that should be carried out depend on the instrument being used.

Magnetic instruments
The correct number of non-magnetic collars should be run in the BHA (see Fig. 7.4). All the collars used should be regularly checked for possible hot spots. The spacing-out of the compass within the collar is also important so that the compass is not placed too near a connection.

The compass unit should be checked against a master compass. MWD directional sensors should be calibrated on a test stand before being sent to the rig. The correct calibration coefficients should be applied at the well site. The correct magnetic declination should be applied. This will vary with time and geographical location.

It should be ensured that neighbouring casings and other sources of magnetic interference are out of range of the magnetic sensors. To check for magnetic interference when drilling beneath a casing shoe both magnetic and gyro instruments can be used and their results compared. When there is agreement between the two sets of results there is no further need to take gyro surveys.

Gyroscope instruments
A conventional (free) gyro relies on an accurate alignment using a telescope sighting onto a benchmark. The benchmark should be precisely located by topographical surveying methods.

Gyro drift should be measured at certain intervals during the survey. The amount of drift will not increase uniformly with time. Once the tool is back on surface, the orientation of the gyro should be checked against the benchmark. If one knows the amount of drift that has occurred during the survey, the observed azimuth can be corrected.

Rate gyros do not require surface alignment, but they do require careful calibration in a test stand. The azimuths and inclinations used on the test stand should be representative of the angles expected in the well.

All gyroscopic instruments are sensitive to excessive vibration and can easily be damaged while running into the hole. Accurate results can only be obtained by careful handling.

Good surveying practices
In addition to carrying out instrument checks, some errors can be eliminated by good operating practices. Where possible, tandem instruments should be run, to provide a check on the reliability of the results and as a back up if one instrument fails to operate properly. Documented records of previous surveys should be kept for all wells. These can be easily accessed if stored on computer files. Plot azimuth and inclination versus depth to identify any anomalies that may have occurred.

When running a multishot tool, continue up beyond the previous casing shoe to give an overlap with the previous multishot and compare results. Subsequent MWD or single shot surveys should be tied back to the definitive survey run to the previous casing shoe.

Keep up to date records of the tools used, serial numbers of non-magnetic collars, survey instruments, etc., so that faults can be traced back and checked out.

Surveying programme
The well must be surveyed at certain intervals to ensure it is following the planned well path. The survey interval will vary depending on which part of the trajectory is being drilled. During kick-off, surveys should be taken at least every 30 ft. In the tangential section when inclination is not changing so rapidly a survey interval of 90 ft or 120 ft may be sufficient. With the introduction of steering tools and MWD the well path can be surveyed almost continuously. The final trajectory is usually measured by running multi-shot tools or the more complex instruments described in Chapter 7.

Estimation of Error in Borehole Position

Although the errors in directional surveying can be substantially reduced by careful planning and operational procedures, the uncertainty in determining borehole position cannot be eliminated completely. In order to quantify the range of this uncertainty, a mathematical model can be constructed to take account of all the contributing sources of error. Most of the errors already discussed can be described as systematic (i.e. they consistently distort the true result in one direction). The effect of a local magnetic field in the drill string, for example, will affect the magnetic compass in the same way at each survey station. Such an error cannot be averaged out over a large number of survey stations as would be the case with random errors. The error model produced by Wolff and de Wardt identified the following errors:

(a) Magnetic compass error, which includes effects such as instrument error, value of magnetic declination and magnetization of the drill string.
(b) Gyroscopic compass error, which includes initial orientation error and gyro drift.

TABLE 9.3 Input of Error Tolerances for Uncertainty Model

	Relative depth, ϵ (10^{-3})	Misalignment ΔI_m (deg.)	True inclination, ΔI_{to} (deg.)	Reference error, ΔC_1 (deg.)	Drill string magn. ΔC_2 (deg.)	Gyro compass, ΔC_3 (deg.)
Good gyro	0.5	0.03	0.2	0.1	—	0.5
Poor gyro	2.0	0.2	0.5	1.0	—	2.5
Good magnetic	1.0	0.1	0.5	1.5	0.25	—
Poor magnetic	2.0	0.3	1.0	1.5	5.0 ± 5.0	—
Weighting	1	1	$\sin I$	$\sin I$	$\sin I \sin A$	$(\cos I)^{-1}$

(After Wolff and De Wardt; courtesy of the SPE.)

(c) Misalignment (true inclination error), which includes the effects of poor centralization and bending effects.

(d) Depth-measurement errors, which include the inaccuracies in wireline or drill pipe measurements.

The model requires an estimate of the size of error related to each of these sources. For each type of survey, therefore, an upper and lower limit must

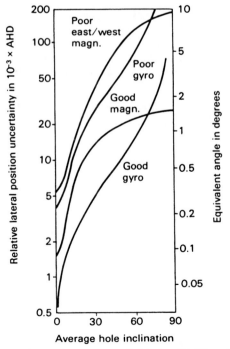

Fig. 9.9. Survey uncertainty using error model (after Wolff and de Wardt; courtesy of the SPE).

be specified. For example, a misalignment error of 0.2° may be given for a poor gyro, while 0.03° may be given for a good gyro. By inputing all the good error estimates, a small ellipsoid is generated to give the optimistic case. This can then be compared with the pessimistic case based on the poor estimates. The model calculates the dimensions of the three orthogonal axes that describe the ellipsoid (i.e. one parallel to the borehole axis due to depth error, one lateral axis due to compass and misalignment error, and the third axis perpendicular to the other two due to the inclination and misalignment error.)

The most significant error is in the lateral positioning of the borehole. This model showed that, by applying the error ranges given in Table 9.3 for an inclined borehole, the magnitude of the lateral error is as shown in Fig. 9.9.

The lateral uncertainty for all types of survey instrument increases with increasing inclination, but not at the same rate. Using these results, therefore, it is possible to estimate the lateral error in bottom hole position (e.g. at 30° inclination with a good gyro the error in position at 10,000 ft is about $0.005 \times 10,000 = 50$ ft. With a poor gyro the error is $0.03 \times 10,000 = 300$ ft).

QUESTIONS

9.1. The following extract is taken from a survey report. Calculate the coordinates of the last three survey stations using the average angle method. The target bearing is 095°. Plot the results on both horizontal and vertical planes.

No.	MD	INC	AZI	N	E	TVD	Vert. Section
15	6000 ft	20°	087°	10 ft	800 ft	5900 ft	796.08 ft
16	6093 ft	22°	091°				
17	6186 ft	23.5°	096°				
18	6279 ft	26°	111°				

9.2. For the last three survey stations in Question 9.1, calculate the dog-leg severity.

9.3. Use (a) tangential, (b) balanced tangential, and (c) minimum curvature methods to calculate the coordinates of station 81 in the following example. The target bearing is 220°.

No.	MD	INC	AZI	N	E	TVD	Vert. Section
80	7000 ft	35°	241°	−3500 ft	−500 ft	5800 ft	3002.85 ft
81	7200 ft	40°	225°				

9.4. The following table gives the coordinates of two adjacent wells drilled from an offshore platform. Estimate the true distance between the wellbores at their closest point.

Well X			Well Y		
TVD	*N*	*E*	*TVD*	*N*	*E*
1207.61	52.31	−3.71	1210.21	52.62	12.32
1236.29	53.67	−2.32	1239.65	53.98	11.05
1264.85	55.02	−1.58	1265.31	56.51	9.12
1291.76	57.35	−1.62	1290.15	58.87	9.62

9.5. Discuss the survey errors associated with
   (a) magnetic instruments;
   (b) gyro instruments.
9.6. According to Fig. 9.9, what is the error in surveying a 5000 foot interval of deviated hole whose average inclination is 20° with (a) a good gyro and (b) a poor gyro?
9.7. An operator wishes to survey a directional well such that the maximum error does not exceed 150 ft at a total depth of 15,000 ft. If the average inclination of the well is 45°, which type of surveying instrument should be used?

FURTHER READING

"A guide to accurate wellbore survey calculations", P. Dailey, *Drilling-DCW*, May 1977.

"Directional survey calculation", J. T. Craig and B. V. Randall, *Petroleum Engineer*, March 1976.

"Borehole position uncertainty. Analysis of measuring methods and derivation of systematic error model", C. J. M. Wolff and J. P. deWardt, S.P.E. paper no. 9223.

"Analysis of uncertainty in directional surveying" J. E. Walstrom, A. A. Brown and R. P. Harvey, *Journal of Petroleum Technology*, April 1969.

"Program challenges directional survey accuracy claims", M. Stephenson, *Oil and Gas Journal*, 20 August 1984.

"An improved method for computing directional surveys", G. J. Wilson, *Journal of Petroleum Technology*, August 1968.

"Synthesis of International Geomagnetic Reference field values". D. R. Barraclough and S. R. C. Martin. *HMSO, Report No. 71/1*.

"Directional Drilling Survey Calculation Methods and Terminology", *API Bulletin D20*, 1985.

"Directional Survey Calculation Methods Compared and Programmed". N. P. Callas, P. C. Novak, J. R. Henderson, *Oil and Gas Journal*, 22 Jan 1979.

"How to get the best results from well surveying data", J. L. Thorogood, *World Oil*, April 1986.

Chapter 10

DRILLING PROBLEMS IN DIRECTIONAL WELLS

Directional wells present a number of drilling problems in addition to those encountered in vertical wells. These additional problems are related to factors such as the well profile and the reduced axial component of gravity acting along the borehole. As the angle of inclination increases, drilling problems become more severe. The particular problems related to highly deviated wells (i.e. those with inclinations over 60°) will be dealt with in the next chapter.

The degree of difficulty in drilling a well is usually reflected in the time taken to complete the well and the total drilling cost. This can be seen in Fig. 10.1, which compares the expected duration of the drilling programme with the actual time taken. This graph clearly shows that having to sidetrack a well means that considerable extra time and money have to be spent. To eliminate the need for sidetracks and other remedial measures, it is vitally important to understand how drilling problems can arise. It should be possible by careful planning to avoid certain problems and thus keep drilling costs down.

CONTROL OVER BOREHOLE TRAJECTORY

A directional well must intersect a target that might be several miles away from the surface location. To reach the target, the wellbore must be directed along a pre-determined trajectory. Careful selection of various bottom hole assemblies is necessary to accomplish this. Many factors may deflect the bit away from the planned wellpath, such as:

(a) formation effects (e.g. bed boundaries);
(b) excessive WOB;
(c) incorrect choice of BHA.

A certain amount of turning either to the left or right is permissible,

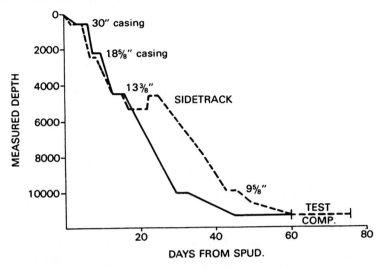

Fig. 10.1. Depth–time graph showing effect of sidetrack.

provided the projected wellpath will still intersect the target zone. The maximum allowable turn can be determined by considering Fig. 10.2.

Point A represents a known survey station on the planned well path. Point T represents the target with a radius R. If ΔH is the horizontal distance between A and T, the maximum deviation away from the planned trajec-

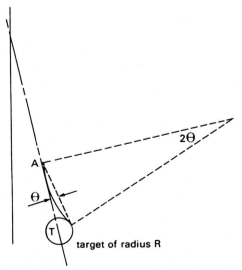

Fig. 10.2. Maximum turn to hit edge of target.

tory which will still allow the well to intersect the target can be found from:

$$\theta = \sin^{-1}\left(\frac{R}{\Delta H}\right)$$

Assuming this change to occur over a smooth arc the amount of turn per 100 ft drilled is given by

$$t = \frac{2\theta}{\Delta H} \times \sin I \times 100$$

where I = angle of inclination. If the actual rate of turning exceeds this value then the wellbore will not pass through the target. Corrective action (e.g. changing the BHA) must be taken to bring the well back on course.

If it is known that formation effects are likely to alter the wellpath, and the extent of this alteration can be reliably predicted, then this can be taken into account in the well plan. If it is known from experience that the bit will tend to walk 10° to the right over a certain interval, then the directional driller can compensate for this by leading the well off 10° to the left. At all times it is important to monitor the wellpath closely. The introduction of MWD in directional drilling has proved to be a major benefit in this respect. The actual wellpath as calculated from surveys while the well is being drilled can be compared with the intended wellpath. Any divergence from the planned course should be recognized and corrective action should be taken. The greater the divergence, the longer the correction run required. It becomes more difficult with increasing depth and inclination to correct the well path. It may be possible to make small corrections by adjusting the position of certain stabilizers in the next rotary BHA. If this is not successful, a downhole motor and bent sub will usually be run.

The basic information required to plan a correction run is the current hole position, current azimuth and inclination, the amount of turning required, and the capability of the deflecting tools available. Any correction run should be made without incurring severe dog-legs. Downhole motors and bent subs are often used to achieve a smooth and gradual turn. The amount of turn a downhole motor and bent sub can achieve can be estimated from manufacturer's tables (Table 10.1). In practice, the formation hardness may affect the rate of turn. A bent sub with a larger offset may be required in harder formations to achieve the required turn. It is therefore better to make correction runs in a softer formation if possible.

Correct orientation of the bent sub is vital, taking into account the effects of reactive torque. Again, the use of MWD or steering tools has an important role to play in this operation. The drill pipe may have to be turned on surface to point the toolface in the required direction. Any torque taken up by the drill string itself must be worked free. The WOB can be adjusted while on bottom to ensure that the toolface reading on the surface readout is at the correct heading. Surveys of inclination and azimuth should also be taken as the hole is drilled. It should be remembered that, since the sensors may be located 40–60 ft behind the bit, the

TABLE 10.1 Amount of Turn Expected from Downhole Motor and Bent Sub

	Deflection expected (degrees/100 ft)									
PDM:	*5 in.*		*6½ in.*		*7¾ in.*		*9⅝ in.*		*12 in.*	
Bent sub angle	*Hole size*	*Deflection angle*	*Hole size*	*Deflection angle*	*Hole size*	*Deflection angle*	*Hole size*	*Deflection angle*	*Hole size*	*Deflection angle*
1°	6″	3°30′	8¾″	2°30′	9⅞″	2°30′	13½″	2°00′	17½″	2°00′
1½°		4°45′		3°30′		3°45′		3°00′		4°00′
2°		5°30′		4°30′		5°00′		4°30′		5°30′
1°	6¾″	3°00′	9⅞″	1°45′	10⅝″	2°00′	15″	1°45′	22″	2°00′
1½°		3°00′		3°30′		2°30′				3°15′
2°		5°00′		3°45′		4°15′		3°45′		4°00′
2½°		5°45′		5°00′		5°30′		5°00′		
1°	7⅞″	2°30′	10⅝″	1°15′	12¼″	1°45′	17½″	1°15′	26″	1°45′
1½°		3°30′		2°00′		2°30′		2°15′		3°00′
2°		4°30′		3°00′		3°30′		3°00′		3°30′
2½°		5°30′		4°00′		5°00′		4°30′		

(Courtesy of Smith International Inc.)

effect of changing the well path may not be seen immediately in the survey results.

Once the correction run has been made, another BHA is run. There is a tendency for the new BHA to follow the curvature of the correction run, and so cause a further divergence from the intended course. To avoid this problem of over-compensation, the orientation should be altered towards the end of the correction run to guide the next BHA into the required heading.

INTERSECTIONS

The spacing of the conductors is often very close when drilling from multiwell platforms. On North Sea platforms, it is quite common for 30-in. conductors to be driven at 7–12 ft between centres. Drilling out beneath the conductor and kicking-off in the proximity of other wells is a very difficult operation which must be carefully controlled. The well being drilled must not be allowed to damage the adjacent wells, which may be high-capacity producers or high-pressure gas injectors. The consequences of a collision beneath a large platform could be serious:

(a) If the casing string of an existing well were damaged, the integrity of that well would be put at risk.
(b) If the tubing string of an existing well were damaged, flowing oil or gas could escape into the formation or cause a blow-out. The safety of the entire platform would then be put at risk, and there would be potential damage to the environment by pollution.

Good planning procedures will reduce the risk of an intersection:

(a) Slot selection should ensure that the optimum slot is allocated for each target, so that wellbores do not cut across each other. Ideally, every target should be matched up with the corresponding slot before development drilling begins. However, this is not always practical since future target locations will be changed as more information becomes available to the geologists. The slot selection process must therefore be flexible, and the engineer planning the wells must keep up to date with the geologists' revised targets. Choosing the most convenient slot for one target may mean another slot becoming completely blocked, making it difficult to use for a future well.
(b) Some separation can be achieved by varying the KOP in adjacent wells. One well might kick off at 2000 ft, while its neighbour is kicked off at 1600 ft (Fig. 10.3). Variations in build-up rate can also be used.
(c) Accurate surveys are essential when trying to drill a new well through a cluster of existing wells. The position of all adjacent wells should be known to a high degree of accuracy. Operating companies are now taking advantage of more sophisticated surveying tools to survey the top section of the wellbore. As the new well is being drilled, the ellipse of uncertainty at each survey depth should be calculated and compared

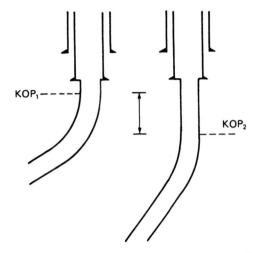

Fig. 10.3. Variation in KOP to increase separation.

to that of the existing wells. If it is clear that the well is coming too close to an existing well, it must be nudged away to a safer position.

Although collisions have been reported, very few major problems have occurred. This is due to the recognition of warning signs before a serious problem has developed.

(a) The cement sheath around the existing cased hole has to be penetrated first. This can be detected by monitoring the cuttings for cement particles.
(b) The driller should recognize changes in certain drilling parameters (torque, vibration) as the bit hits cement or steel casing.
(c) Since both wells will be vertical, or near vertical, there will be a tendency for the well being drilled to "bounce off" the existing well.

In spite of these warning signs, extreme care must be taken when drilling through dense clusters of wells. Many operators have a policy laid down to deal with a potential collision. If the ellipses of uncertainty come within a certain distance, or if they overlap, the adjacent well must be shut in; a wire-line plug is run into the tubing and the well is de-pressurized. The casing annulus ($9\frac{5}{8}$ in. × $13\frac{3}{8}$ in.) of the existing well is pressured up to 1000 psi and monitored while the other well is being drilled. Only after the critical point has been safely passed and the wells are shown to be diverging can the producing well be opened up again.

DOG-LEG SEVERITY

A dog-leg is an abrupt change in hole angle or direction that causes a sharp bend in the wellpath. It can be detected by increased torque and drag on the drill string. The size of the dog-leg can be calculated from survey results over a 60–90 ft interval using the following equations:

$$\phi = \cos^{-1}[\cos \alpha_1 \cos \alpha_2 + \sin \alpha_1 \sin \alpha_2 \cos(\beta_2 - \beta_1)]$$

or

$$\phi = \cos^{-1}\{\cos(\alpha_2 - \alpha_1) - \sin \alpha_1 \sin \alpha_2[1 - \cos(\beta_2 - \beta_1)]\}$$

where ϕ = dog-leg angle (degrees)
α_1, β_1 = inclination and azimuth at station 1
α_2, β_2 = inclination and azimuth at station 2.

The severity of the dog-leg is expressed as the change in angle per 100 ft drilled, i.e.

$$\text{DLS} = 100 \frac{\phi}{L}$$

where DLS = dog-leg severity (degrees/100 ft)
ϕ = dog-leg angle
L = course length between stations.

Severe dog-legs have a detrimental effect on drilling operations.

Increased Wear on the Drill String and Casing

If there is a dog-leg in the hole, the drill pipe will be subjected to bending stresses. On the inside of the bend, the wall of the pipe will be in compression, on the opposite wall, there will be tension. These effects will be reversed as the pipe rotates through 180°. The pipe therefore undergoes cyclic loading that encourages fatigue and reduces the operational life of the drill pipe.

To limit the bending stresses on the pipe, a maximum permissible dog-leg severity C can be calculated as follows:

$$C = \frac{432{,}000\sigma_b \tanh(KL)}{\pi EDKL} \qquad \text{where} \qquad K = \left(\frac{T}{EI}\right)^{1/2}$$

where C = maximum dog-leg severity (degrees/100 ft)
σ_b = maximum permissible bending stress (psi)
E = modulus of elasticity (psi)
D = outside diameter of drill pipe (in.)
L = half distance between tool joints (L = 180 in. for range 2 pipe)
T = tension loading below the dog-leg (lb)
I = moment of inertia (in.^{4});
for circular pipe, $I = (\pi/64)(D^4 - d^4)$
d = inside diameter of drill pipe (in.).

The value of σ_b depends on the grade of pipe used and the tensile stress σ_t exerted on the pipe ($\sigma_t = T/A$, where A = cross-sectional area of the pipe). For Grade E drill pipe with a tensile stress less than 67,000 psi, the following formula can be used:

$$\sigma_b = 19,500 - \frac{10}{67}\sigma_t - \frac{0.6}{(670)^2}(\sigma_t - 33,500)^2$$

EXAMPLE 10.1

A dog-leg is expected at a depth of 5000 ft. A string of 5-in., 19.5 lb/ft, grade E drill pipe is being used. If the tensile load below this point is 200,000 lb, calculate the maximum dog-leg that can be allowed.

$$\sigma_t = \frac{T}{A} = \frac{200,000}{\frac{1}{4}\pi(5^2 - 4.276^2)} = 38,000 \text{ psi}$$

thus

$$\sigma_b = 19,500 - \frac{10}{67}(38,000) - \frac{0.6}{(670)^2}(38,000 - 33,500)^2$$

$$= 13,800 \text{ psi}$$

$$K = \left(\frac{T}{EI}\right)^{1/2} = \left(\frac{64 \times 200,000}{30 \times 10^6 \times \pi(5^4 - 4.276^4)}\right)^{1/2} = 0.0216$$

$$C = \frac{432,000 \times 13,800 \times \tanh(0.0216 \times 180)}{\pi \times 30 \times 10^6 \times 5 \times (0.0216 \times 180)}$$

$$C = 3.25° \text{ per 100 ft}$$

If the dog-leg severity at this depth exceeds 3.25°/100 ft, the pipe will begin to suffer fatigue damage. The effect of fatigue on the life of the drill pipe can be calculated in terms of the percentage by which the operational life has been reduced. The fatigue damage is a function of the pipe grade, outside diameter, dog-leg severity and tensile load. Hansford and Lubinski produced graphs to determine the percentage of fatigue life expended in drilling a 30 ft interval (Fig. 10.4). These graphs were drawn for both corrosive and non-corrosive environments, and are based on a penetration rate of 10 ft/h and a rotational speed of 100 rpm. Since fatigue life expended is directly proportional to rpm, but inversely proportional to ROP, these charts can be used for any combination of ROP and rpm.

EXAMPLE 10.2

A dog-leg of 3.6° is calculated over a distance of 60 ft. Below this point there is 6000 ft of 5 in., grade E drill pipe. If the ROP is 25 ft/hr and speed is 80 rpm, calculate the reduction in fatigue life, assuming a corrosive environment.

$$\text{DLS} = 100\frac{3.6}{60} = 6° \text{ per 100 ft}$$

From Fig. 10.3 the percentage fatigue life expended is 38% over a 30 ft

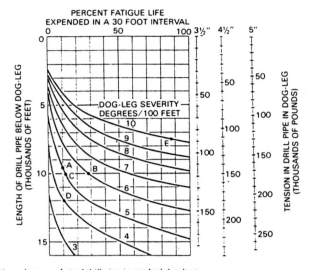

Fatigue damage of steel drill pipe in gradual dog-legs
(non-corrosive environment: drill pipe, 3½, 4½ and
5 in.grade E steel; rotary speed, 100 rpm; drilling rate, 10ft/hr).

NON-CORROSIVE ENVIRONMENT

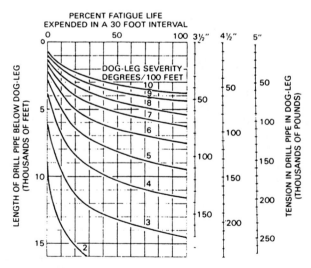

Fatigue damage of steel drill pipe in gradual dog-legs
(extreme corrosion fatigue: drill pipe, 3½, 4½ and
5 in.grade E steel; rotary speed, 100 rpm; drilling rate, 10ft/hr)

CORROSIVE ENVIRONMENT

Fig. 10.4. Effect of dog-leg severity on fatigue life. (a) Non-corrosive environment; (b) Corrosive environment. (Drill pipe 3½ in., 4½ in. and 5 in. grade E steel;
rotary speed 100 rpm; drilling rate 10 ft/h.) (After Hansford and Lubinski;
courtesy of the SPE).

interval. Over the 60 ft interval, and taking account of the actual drilling parameters, this corresponds to:

$$38 \times \frac{60}{30} \times \frac{80}{100} \times \frac{10}{25} = 24\%$$

Therefore, 24% of the fatigue life is expended in drilling through this 60 ft section of hole. This calculation can be repeated for any other dog-legs in the hole, and the effect is cumulative. This procedure gives a guide as to when a particular section of pipe should be replaced.

One practical measure that can be taken to reduce the effect of dog-leg severity on drill pipe is the use of rubber protectors. These are easily attached to the drill pipe and reduce the amount of contact between the drill pipe and the borehole. In a dog-leg, the protectors distribute the bending stresses more evenly over the drill pipe section and so reduce drill pipe fatigue.

Severe dog legs can also exert side forces on tool joints, which, if excessive (over 2000 lb), can cause keyseats and tool joint failures. The maximum dog-leg severity that can be tolerated is given by:

$$C = \frac{108,000F}{\pi L T}$$

where F = lateral force on tool joint (lb)
L = half-length of drill pipe joint (180 in.)
T = tension load (lb)
C = dog-leg severity (degrees per 100 ft).

EXAMPLE 10.3

For a lateral load of 2000 lb and a tensile load of 200,000 lb, what is the maximum permissible dog-leg?

$$C = \frac{108,000 \times 2000}{\pi \times 180 \times 200,000} = 1.9° \text{ per } 100 \text{ ft}$$

Any dog-leg greater than this figure is likely to cause a tool joint failure under these conditions. Drill collar connections can also fail owing to lateral loading caused by severe dog-legs.

Any severe dog-legs in the hole must be considered when designing casing string, since the bending effect will reduce the collapse resistance of the casing. The reduction in collapse resistance is not usually significant unless the dog-leg severity exceeds 10° per 100 ft.

The maximum fibre stress in the casing, S_b, can be determined from

$$S_b = \frac{DE(\text{DLS})}{137,150}$$

where S_b = fibre stress (psi)
D = outside diameter of casing (in.)
E = modulus of elasticity (psi)
DLS = dog-leg severity (degrees per 100 ft).

The maximum fibre stress must not exceed the yield stress of the casing. The effective tension due to bending is calculated by multiplying S_b by the cross-sectional area of the casing, A_t. The equivalent tensile load is therefore given by

$$\text{ETL} = (S_b \times A_t) + T$$

where ETL = equivalent tensile load (lb)
      T = axial tensile load (including buoyancy).

A factor S_2 can be defined as

$$S_2 = \frac{\text{ETL}}{\text{pipe body yield strength}}$$

The factor S_2 is then used to calculate S_1:

$$S_1 = 0.5[-S_2 + (4 - 3S_2^2)^{1/2}]$$

The de-rated collapse resistance is then found by multiplying the full collapse resistance by S_1. If this de-rated figure is less than the applied collapse loading at that depth, the casing of that interval must be re-designed. Some operators select a heavier weight of casing for any section of the hole that has a dog-leg, to allow for extra drill pipe/casing wear at that point.

KEYSEATING

When the drill string is rotating, the drill pipe is kept in tension by the weight of the drill collars. If the pipe has to pass through a severe dog-leg, the pipe will make contact with the side of the hole. Continued drilling in this position will gradually wear away a small-diameter groove in the side of the borehole wall. A cross section taken through the borehole at this point will show that the hole has been enlarged slightly to form a "keyseat" (Fig. 10.5).

The problem arises when tripping out of the hole. The drill pipe may be able to pass through the keyseat, but the larger-diameter drill collars will probably become stuck at the narrow groove. The pipe may be lowered and rotated, but upward movement is prevented by the keyseat.

To free the pipe the keyseat must be reamed out by a string stabilizer or keyseat wiper (Fig. 10.6). The keyseat wiper is normally installed on top of the drill collars. A key seat wiper consists of a mandrel and a sleeve, which acts like a welded blade stabilizer. The blades are $\frac{1}{8}''$–$\frac{1}{4}''$ larger than the diameter of the drill collars. A clutch mechanism ensures that the sleeve engages with the mandrel when pulling out of the hole. If the drill string is rotated and the keyseat reamed out in stages, the drill collars will be able to pass through. It is good practice to install a keyseat wiper in the BHA when drilling directional wells in which dog-legs can be expected.

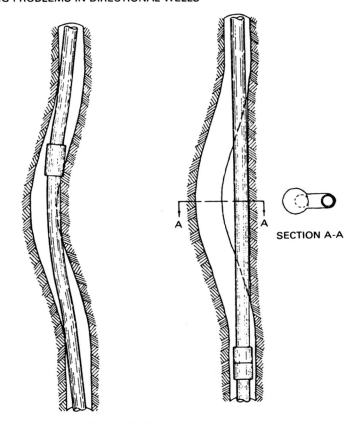

Fig. 10.5. Formation of a keyseat.

WELLBORE INSTABILITY

One of the major causes of stuck pipe is the tendency for some formations to become unstable either during drilling or at some time later. This may cause fragments of rock to fall into the hole and become wedged around the drill collars or bit. Borehole instability is more likely to occur where the following conditions exist (Fig. 10.7):

(a) shale zones containing a high percentage of swelling clays (sodium montmorillonite);
(b) steeply dipping or fractured formations (limestone);
(c) overpressured shale zones;
(d) turbulent flow in the annulus causing washouts (erosion) in soft formations.

It becomes more difficult to lift these fragments out of the hole as the inclination increases.

Fig. 10.6. Use of a keyseat wiper to ream out a keyseat (courtesy of Grant Tool Co.).

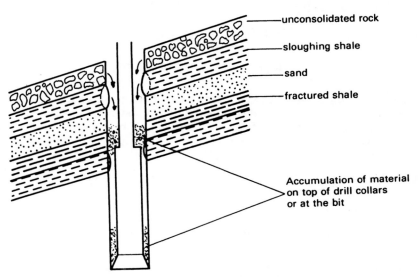

Fig. 10.7. Borehole instability.

Most of the problems can be related to shale zones. Most shales will absorb water to some extent, lowering the compressive strength of the rock and allowing it to expand. In brittle shales, the water is absorbed along fracture planes, weakening the structure and allowing large fragments to fall into the hole. The degree of swelling depends on the composition of the shale and its moisture content. Some shales undergo plastic deformation on contact with water; the shale expands into the borehole and seals off the annulus. Mud rings can form in the annulus that will block the flowline and prevent circulation.

The swelling of the shale can be reduced by using inhibited drilling muds. Muds containing potassium or calcium salts (e.g. KCl or gypsum) will reduce the adsorption of water onto the clay platelets. Oil-based muds also act as inhibitors. The mud properties must be carefully selected to combat this problem (e.g. low fluid loss, low solids content). The mud weight is also important, since the hydrostatic pressure may help to support these unstable formations. If the mud weight is too high with respect to the formation pore pressure, however, there is a risk of differential sticking in permeable zones.

DIFFERENTIAL STICKING

To prevent the flow of formation fluids into the wellbore, the hydrostatic mud pressure in the borehole must balance or exceed the pore pressure. In a permeable zone, a natural filtration process will take place whereby the fluid content of the mud will invade the formation while the solids will build up on the wall of the borehole to form a filter cake. If the filter cake becomes thick, the drill collars may come into contact with it and become embedded. If the positive pressure differential is large (about 1000 psi), it may be difficult to free the pipe (Fig. 10.7). This is a problem frequently encountered in directional wells, since the drill collars will tend to sag on the Low Side of the hole and become stuck against the filter cake. The risk of differential sticking is increased if the pipe is allowed to stay static for a period of time.

The force required to free the pipe can be expressed as

$$F = \Delta p A_c f$$

where Δp = differential pressure (psi)
A_c = area of contact between pipe and hole (in.^{2})
f = friction coefficient between pipe and hole.

By reducing each of the three contributing factors the problem may be solved.

(a) Differential pressure can be reduced by lowering the hydrostatic mud pressure. Water or a lighter weight mud (diesel) could be circulated down the drill string. However, this could lead to a kick in another zone within the open hole section.

(b) The contact area can be reduced by using spiral drill collars with wide grooves cut into their external profile. These collars are recommended for directional drilling when stuck pipe is likely. Heavy weight drill pipe may also be used.
(c) The friction coefficient can be reduced by using lubricants or oil-based muds.

Since the risk of differential sticking is increased by keeping the pipe stationary, reciprocation of the pipe is necessary while drilling is stopped.

FREEING STUCK PIPE

The method adopted to free the pipe will depend to some extent on how the pipe became stuck in the first place. In most cases this can be deduced from knowing whether it is possible to circulate, reciprocate or rotate the pipe.

(a) If the hole has been bridged by rock fragments, only limited reciprocation will be possible. In such cases it is very difficult to free the pipe without resorting to back-off methods.
(b) If the pipe is stuck in a keyseat, it will be possible to go down, rotate and circulate. The only effective means of freeing the pipe is to ream the keyseat out.
(c) Differential sticking is recognized by being able to circulate but not to rotate or reciprocate. For differential sticking, a special soaking fluid can be pumped down the drill string and displaced into the annulus adjacent to the stuck zone. If time is allowed for the fluid to soak into the interface between the collars and the borehole, while the pipe is worked from surface, the pipe may become free. This method requires careful volume calculations and requires time to allow the soaking fluid to work, but it has been used successfully in many wells.

If no progress has been made in freeing the pipe after several days, a decision has to be made. There are several options facing the operator:

(a) continue trying to free the pipe;
(b) back-off the drill string, and fish out the remaining part of the string;
(c) back-off the drill string, and sidetrack further up;
(d) back-off the drill string, and abandon the well.

The final choice will depend on individual circumstances (e.g. how important is this hole, what is the probability of recovering the fish, will it be possible to drill a sidetrack). Inevitably, the experience gained in similar previous circumstances will influence the decision. The operator will also have to consider the cost and the consequences of each option.

BACKING-OFF THE DRILL STRING

Backing-off is the term used to describe the method of disconnecting the drill string at some depth above the stuck point. The pipe must be

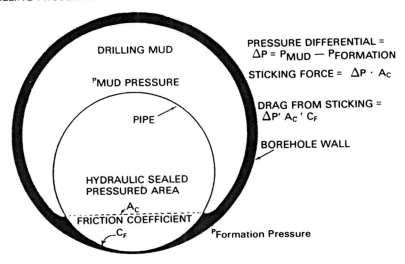

Fig. 10.8. Differential sticking.

backed-off at the deepest point that is practical to allow maximum recovery. Once the upper section has been removed, the lower section must be fished out. The top of the fish should therefore be easily accessible to fishing tools. For this reason, the back-off point may be selected at a tool joint or collar in a section of the hole that has no washouts. One or two free joints of pipe may be left in the hole above the stuck point to allow easier washover operations.

Location

The first stage in performing a back-off is to locate the point at which the string is stuck. There are two methods of doing this.

Pipe stretch
Pick up the weight of the drill string, and mark the drill pipe at the rotary table. Pull an additional amount P and make another mark. Repeat the test and compare results to obtain the elongation of the free section of pipe, e. The length of the free section of pipe can then be calculated from

$$L = \frac{EeW}{40.8P}$$

where L = length of free pipe (ft)
 E = modulus of elasticity (psi)
 P = differential pull (lb)
 e = pipe stretch (in.)
 W = weight of pipe (lb/ft).

This method gives a rough guide as to where the pipe is stuck, but it is not very accurate.

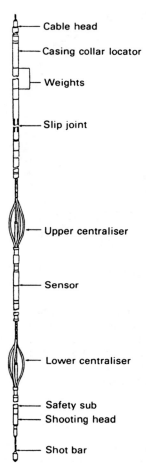

Fig. 10.9. Free point indicator.

Free point indicator
The free point indicator (Fig. 10.9) is an electronic device run on wireline through the inside of the drill string. The key element consists of a sensitive strain gauge that can measure stretch and torque movements in the drill string. When the tool is set at any particular depth, the strain gauge is attached to the inside of the drill string. When the drill string is pulled or rotated from surface, the reading on the strain gauge will indicate how free the pipe is at that point. The reading in a free section of pipe will be much higher than in stuck sections. The stuck point can be determined by interpreting the strain gauge readings over a number of successive depths.

Parting the Drill String

The drill string is parted by an explosive charge positioned at the back-off point. The charge is run on the same assembly as the free point indicator and a CCL (casing collar locator) to help place the charge opposite a connection at the required depth.

The connections in the drill string are tightened by applying right-hand torque from surface. To make it easier for the pipe to be disconnected, left-hand torque is then applied so that the upper section will spin free when the charge is detonated. Having successfully backed-off the pipe, circulate to clean up the top of the fish before pulling out of the hole.

FISHING

The first stage in carrying out a fishing job is to collect all the relevant information concerning the fish. It is useful to make a sketch showing the dimensions of the various components and the depth at which the fish is located. When the free section of pipe has been recovered, its severed end should be inspected for any irregularities that may also be on the top of the fish. If the profile of the top of the fish is not certain, an impression block

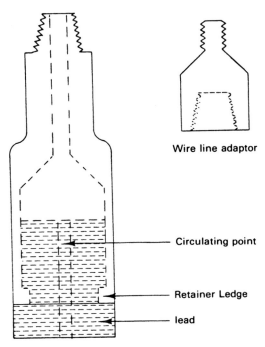

Wire line adaptor

Circulating point

Retainer Ledge

lead

Fig. 10.10. Impression block.

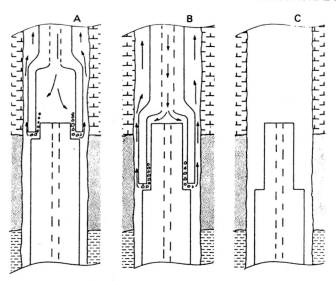

Fig. 10.11. Milling a fishing neck with a throated mill dressed with cutting material on the bottom and inside.

can be run (Fig. 10.10). This tool consists of a steel body with a moulded block of lead that is soft enough to carry an imprint of the fish. The block has a circulating port so that fluid can be pumped through it to clear away any debris lying on top of the fish.

Milling

Milling tools can be used to dress the top of the fish to allow an overshot a better chance of latching onto it (Fig. 10.11). Milling is carried out at low rpm to ensure that the mill does not slip off the fish. The cutting structure is inlaid with tungsten carbide to increase wear resistance.

Overshot Tools

Overshot tools offer the best chance of recovery by latching onto the external profile of the fish. This is only possible if there is sufficient clearance between the fish and the borehole. An overshot can be made up to latch onto various different sizes of tubing, drill pipe, tool joints or drill collars. The major components of an overshot are shown in Fig. 10.12:

(a) a guide to help the overshot get over the fish;
(b) a bowl to house the gripping elements;
(c) a top sub for connecting the overshot to the running string.

The bowl is designed with a tapered section that accommodates the grapple

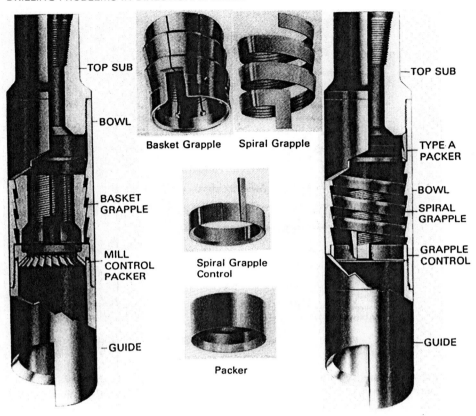

Fig. 10.12. Elements of an overshot (courtesy of Bowen Tools).

(spiral or basket grapples may be fitted). The overshot is rotated clockwise as it is lowered down over the fish. The grapple will expand, allowing the top of the fish to enter the bowl. When rotation is stopped, an upward pull is exerted and the tapered section of the bowl forces the grapple to contract and secure the fish. If it is necessary at any time to release the fish, a sharp downward blow followed by rotation will break the hold.

The overshot also has a pack-off element which seals against the bowl. The pack-off has an internal lip that seals around the fish (Fig. 10.13). By pumping fluid down through the overshot the fish can be circulated clean to assist in the recovery.

Typical Fishing Assembly

A typical fishing assembly is shown in Fig. 10.14, for use with an overshot. A safety joint located above the overshot provides a quick release should

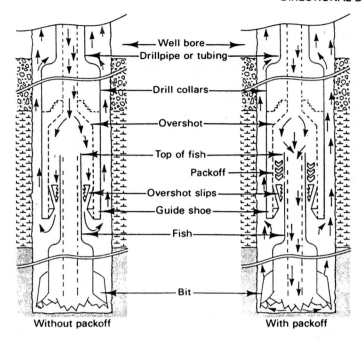

Fig. 10.13. Circulating through fish with pack-off.

the entire fishing string become stuck. A bumper sub is a kind of slip joint used to deliver a sharp downward blow to disconnect the overshot from the fish.

A hydraulic jar is used to apply a heavy upward blow to help jerk the fish free. The tool consists of a mandrel that is allowed to slide inside a cylinder. As the mandrel is pulled up, it is restrained by a pressure differential within the cylinder. At a certain point, the restraining mechanism is released and the mandrel delivers a sharp upward blow against the cylinder. The upward force can be increased by adding heavy drill collars on top of the jars. A further component, known as an accelerator, contains high-pressure gas (nitrogen), which is released when the jars operate. To re-set the jars, the string is lowered and the process is repeated. As many as ten blows per minute may be required to exert enough force to free the fish.

SIDETRACKING

If repeated attempts at fishing prove unsuccessful, the components of the BHA still stuck in the hole must be abandoned. The fishing assembly is released and pulled out of the hole. A cement plug is set above the fish so that a sidetrack may be drilled to reach the target.

A caliper log should be run to calculate the volume of cement required.

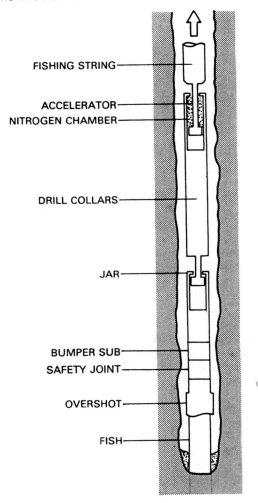

Fig. 10.14. Typical fishing assembly.

The plug should be at least 200 ft long above the fish. An excess volume should also be included to allow for contaminated cement to be drilled out by the bit. The cement should have a high density and be pumped through open-ended drill pipe. The cement must then be allowed to harden. The strength of the plug is determined by the drill bit. A constant ROP of around 1 ft/hour indicates a strong homogeneous cement plug.

The sidetracking assembly will normally consist of a bit, downhole motor, bent sub, MWD tool and drill collars. The choice of bit depends on the formation to be drilled and it should be able to complete the initial sidetrack without having to be replaced. A short PDM should be selected, together with a bent sub that will provide the necessary change in the well path. The most popular size of bent sub is 1–2°, but in harder formations more offset may be required. An MWD tool contained in non-magnetic drill collars will provide good directional monitoring and save time in re-orienting the bent sub when necessary. The remainder of the assembly is made up with drill collars to provide enough WOB.

In planning the trajectory of the sidetrack it is important to ensure that the bit is directed towards the target and that it avoids the fish. It is generally easier to drill out the Low Side of the hole, but this may require building more inclination later on as the hole approaches the target. Sidetracks usually involve both a change in direction and in inclination. The trajectory should be planned to achieve a gradual change so that severe dog-legs do not occur.

QUESTIONS

10.1. In a directional well in which the present inclination is 20° and the azimuth is 035° a sidetrack must be drilled to drop the inclination by 2° and turn the hole 40° to the right. This correction is to be made in 12¼-in. hole over the next 300 ft using a 7¾-in. downhole motor and bent sub.
   (a) Calculate the expected dog-leg severity.
   (b) From Table 10.1 select the correct size of bent sub required.
   (c) Allowing 30° of reactive torque, determine the toolface the directional driller should use.

10.2. Calculate the maximum permissible dog-leg severity for a string of 4½-in., 16.6 lb/ft grade E drill pipe. Assume a tensile load of 180,000 lb. (ID of pipe = 3.826 in.)

10.3. From survey data a dog-leg of 1.3° has been calculated over a course length of 30 ft. Below this point there is a tension load of 150,000 lb due to the weight of a string of 4½-in. grade E drill pipe. The penetration rate is 10 ft/hr and the rotary speed is 100 rpm. Calculate the reduction in fatigue life due to this dog leg for
   (a) a non-corrosive environment;
   (b) a corrosive environment.

10.4. Repeat Question 10.3 for a penetration rate of 30 ft/hr and a rotary speed of 70 rpm.

10.5. In a directional well, the maximum dog leg severity is 5° per 100 ft. For a tensile load of 100,000 lb what is the maximum side force that can be tolerated to prevent tool joint failure?

10.6. The drill string becomes stuck while drilling a directional well. The driller reports that he can still circulate, but cannot rotate or reciprocate the pipe.
   (a) What is the most likely cause of the stuck pipe?
   (b) What action should the driller take to free the pipe?

10.7. Describe how an overshot may be used to recover a section of pipe that has been left at the bottom of the hole.

FURTHER READING

"Oil mud aids in reducing problems and cost of North Sea platform development drilling", B. J. Holder, S.P.E. paper no. 8160.

"Drilling, evaluating and completing high angled wells in the North Sea", N. Woodall-Mason, S.P.E. paper no. 8161.

"Conoco cuts North Sea drilling time by 40%", J. Shute and G. Alldredge, *World Oil*, July 1982.

"Cumulative fatigue damage of drill pipe in dog legs", J. E. Hansford and A. Lubinski, *Journal of Petroleum Technology*, March 1966.

"Maximum permissible dog legs in rotary boreholes", A. Lubinski, *Journal of Petroleum Technology*, February 1961.

"Top drive drilling for deviated wells", *The Oilman*, April 1984.

"How to free a stuck string", M. W. Aulenbacher, *Drilling–DCW*, June 1976.

"Spot fluid quickly to free differentially stuck pipe", H. D. Outmans, *Oil and Gas Journal*, 15 June 1974.

"Casing Wear: Laboratory Measurements and Field Predictions", J. P. White and R. Dawson, S.P.E. paper no. 14325.

Chapter 11

HIGHLY DEVIATED AND HORIZONTAL WELLS

The tools and techniques discussed in earlier chapters are normally used to drill directional wells whose maximum inclination is about 60°. Highly deviated wells may be described as those wells whose inclination exceeds 60° for most of their length. It is possible to extend directional drilling techniques to increase the inclination to 60–90°, although alterations may have to be made to drilling practices. Modifications to standard rig equipment may also be necessary to drill these high-angled wells successfully. A horizontal well may be defined as a well which is drilled to an inclination of 90°, and maintains this inclination for a significant distance. Owing to the need for special equipment and the longer drilling times that must be expected, horizontal wells are considerably more expensive than conventional deviated wells.

The most obvious advantage to be gained by extended reach drilling is the increased in horizontal displacement from a central platform. Consider the example shown in Fig. 11.1. Taking KOP = 2000 ft, build up rate = 2°/100 ft, target TVD = 10,000 ft and inclination = 60°, the horizontal reach is 10,992 ft, corresponding to a drainage area of 13.6 square miles. By increasing the inclination to 80°, a similar well would achieve a horizontal displacement of 31,737 ft, corresponding to a drainage area of 113.5 square miles. Increasing the inclination by 20°, therefore, allows the horizontal reach to increase by a factor of almost 3, and the drainage area to increase by a factor of more than 8. These longer reach wells can reduce the number of platforms required to exploit the reserves in offshore areas (Fig. 11.2).

Another potential benefit of drilling highly deviated wells is the increased length of the completion zone through the reservoir. Assuming the formation is horizontal, an 80° wellbore has almost three times the penetration through the reservoir than that of a wellbore at 60° inclination (Fig. 11.3). This allows much more of the reservoir to contribute to the well's productivity. In some oilfields a small number of horizontal wells could drain the reservoir much more efficiently than a greater number of conventional wells.

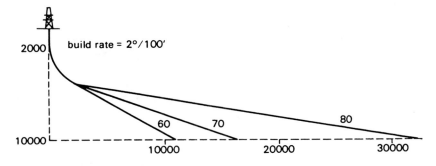

Fig. 11.1. Extended reach of highly deviated wells.

Inclination (deg)	Horizontal reach (ft)	Drainage area (square miles)
60	10,992	13.6
70	16,469	30.6
80	31,737	113.5

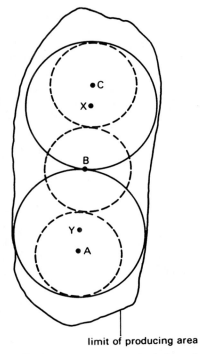

limit of producing area

Fig. 11.2. Reducing number of platforms by extended reach drilling. (Platforms A, B and C being replaced by platforms X and Y.)

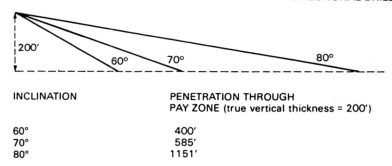

INCLINATION	PENETRATION THROUGH PAY ZONE (true vertical thickness = 200')	
60°	400'	
70°	585'	
80°	1151'	

Fig. 11.3. Increased penetration of reservoir using highly deviated wells.

The potential benefit of increased production has led to a renewed interest in horizontal drilling over recent years. Horizontal drilling dates back to the late 1800s, but the major advances occurred in the 1950s through the efforts of John Eastman and others. The Russians were also involved in some pioneering work. Although these early efforts were technically successful, the relatively low price of oil did not justify the extra costs involved. When oil prices rose in the late 1970s, several companies began examining the possibility of using these techniques.

The major technical problems in drilling highly deviated wells and horizontal wells are related to the effect of gravitational forces as the angle of inclination increases. For directional wells with inclinations of 20–40°, the component of gravity acting along the axis of the borehole is sufficient to allow tools to be run easily into the hole and exert enough WOB. As the angle of the hole increases, the axial component reduces while the lateral component increases. The result of this is:

(a) to increase frictional resistance against the borehole, making it more difficult to run and pull tools;
(b) to increase the tendency for solids to settle out from the drilling fluid and cement slurries;
(c) to make it more difficult to control direction and apply WOB.

In principle, all these problems can be overcome, but only at much higher costs. Surveys from recent horizontal wells show that the cost may be 2–5 times greater than for a comparable vertical well. Horizontal wells will therefore only be drilled where the operator expects a substantial benefit in terms of increased productivity.

Although horizontal wells can provide a solution to many reservoir development problems, their application is not universal. A horizontal well will not be effective in a layered reservoir where vertical permeability is low. If the producing interval is very thin, a horizontal well may easily miss the target. The technical problems, the potential benefits and the additional risk involved must all be considered before starting a horizontal drilling programme.

APPLICATIONS OF HORIZONTAL WELLS

Increased Production from a Single Well

The greater contact area of the wellbore through the producing zone allows a much longer completion interval than would be possible in a less deviated well. With more of the formation contributing directly to the production, higher flow rates can be expected.

The productivity index PI for a vertical well is given by:

$$PI = \frac{7.08 \times 10^{-3} \times kh}{\mu B_0 \ln(r_e/r_w)}$$

where PI = productivity index (bbl/day/psi)
     k = permeability (millidarcies)
     h = reservoir thickness (ft)
     μ = viscosity (centipoise)
     B_0 = oil formation volume factor
     r_e = drainage radius (ft)
     r_w = wellbore radius (ft).

The productivity is therefore proportional to the product kh, sometimes referred to as transmissibility. The productivity index for a horizontal well can be approximated by

$$PI = \frac{7.08 \times 10^{-3} \times kL}{\mu B_0} \cdot \frac{1}{\dfrac{L}{h} \ln\left(\dfrac{1 + [1 - (L/2r_e)^2]^{1/2}}{L/2r_e}\right) + \ln\left(\dfrac{h}{2\pi r_w}\right)}$$

where L = length of horizontal section (ft).

The productivity index for a horizontal well may be five times that of a conventional well. Horizontal wells are therefore suited to relatively thin beds that cover a large area, or to formations where the permeability is low. Horizontal drilling can be used as an alternative to hydraulic fracturing as a means of improving production rates from tight formations. Horizontal wells may also be used to improve water injection as a means of improving oil recovery from the reservoir.

Reduction in Coning Problems

When a vertical well is drilled through a relatively thin pay zone overlying an aquifer, there is a tendency for the water to be drawn up into the perforated interval if the vertical permeability is high. This is known as water coning, and leads to an increased water cut in the producing wells.

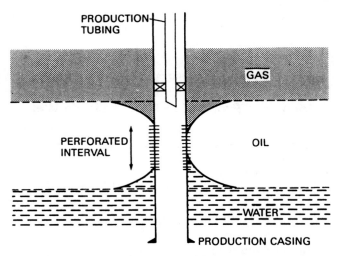

Fig. 11.4. Effects of water coning and gas coning in a vertical well.

This unwanted water production can be reduced by cementing off the lower perforations, and re-perforating higher up. This requires shutting in the well and carrying out a workover. A similar problem exists in the case of a formation that has an overlying gas cap, where the gas is drawn down to the upper perforations (Fig. 11.4).

A horizontal well can alleviate both problems, since it can be strategically placed away from both gas- and water-bearing zones. Owing to the longer length of the completion, the drawdown in the reservoir pressure around the wellbore will also be reduced, giving greater oil recovery before the onset of coning problems.

Intersection of Vertical Fractures

Many reservoirs contain fractures that are vertical or near-vertical at depths greater than 2000–3000 ft. Although the matrix of the rock may be fairly impermeable, the oil may still be able to flow along the fractures. It has been found that in some reservoirs (e.g. fractured limestone) the most efficient way of producing the oil is to drill highly deviated or horizontal wells to intersect as many fractures as possible. If the orientation of the fractures is known, a horizontal well can be planned to intersect the fractures at right-angles.

Enhanced Oil Recovery

Large deposits of highly viscous oil occur in many parts of the world. Since these reservoirs cannot be exploited by conventional means, special techniques have had to be applied, such as the injection of steam or polymers to improve the mobility of the oil. At Cold Lake, in Western Canada, one such project is aimed at recovering oil from a large bitumen deposit.

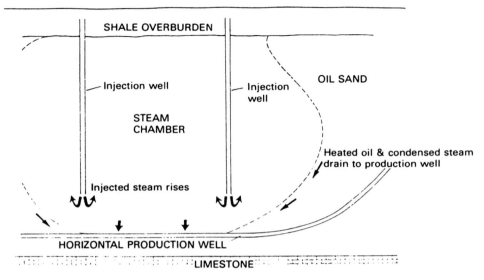

Fig. 11.5. Use of a horizontal production well in an enhanced oil recovery (EOR) project.

Surface mining is not feasible since the deposit is located at a depth of over 1000 ft. It was decided to drill a horizontal well near the base of the deposit to act as a producing well. A number of vertical wells were drilled above this well to allow steam injection into the formation. The viscosity of the oil in the vicinity of the steam injectors was reduced, and it drained downwards under gravity towards the horizontal producing well (Fig. 11.5).

Reducing the Number of Wells and Platforms Required to Develop an Offshore Field

The increased productivity of horizontal wells may result in fewer wells having to be drilled to develop an offshore field. Although the cost of an individual well might be more expensive, owing to the higher inclination, the overall economics of the project could be improved. In shallow reservoirs that cover a wide area the extended reach of horizontal wells may also mean that fewer platforms are necessary. Horizontal drilling may therefore allow the development of a field that would otherwise be considered uneconomic. In offshore fields where there are large distances between bottom hole locations, horizontal wells may be drilled as an alternative to infill drilling, thus further reducing the number of wells to be drilled.

Development of Non-petroleum Resources

Coal seams in certain areas of the world contain large volumes of methane gas. The gas has to be drained off before the coal can be extracted, since a

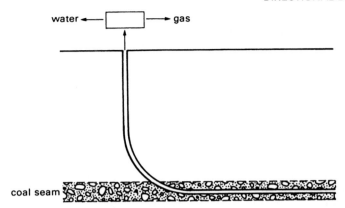

Fig. 11.6. Horizontal well used for methane drainage.

concentration of 5–15% methane in air forms an explosive mixture. To remove the methane, a small-diameter horizontal hole can be drilled through the coal seam. Very close directional control is required, since the coal seam may only be a few feet thick. The orientation of the horizontal drainhole can also be planned to coincide with the direction of maximum permeability through the coal seam (Fig. 11.6).

Apart from the safety aspect, the produced methane is a valuable energy resource. It has been estimated that in the USA alone there is about 400×10^{12} ft^3 of gas contained in underground coal seams.

For coal deposits located at depths beyond conventional mining methods, *in situ* gasification may be used to exploit the reserves. Highly deviated and horizontal wells provide a network of channels for the injection of air and oxygen and for the production of the gas.

OPERATIONAL PROBLEMS RELATED TO HORIZONTAL WELLS

The drilling and completion of highly deviated and horizontal wells introduces many problems that are not encountered, or are not so severe, at lower inclinations.

Drilling Problems

Exerting WOB and running in tools

In normal drilling operations the weight of the drill collars is sufficient to drive the bit and maintain good ROP. As the inclination of the well increases, a larger proportion of the weight is applied to the side of the borehole, and the axial component exerting weight on the bit is reduced. The driving force could be increased by adding more drill collars, but this would only lead to more lateral force, causing increased friction and drag.

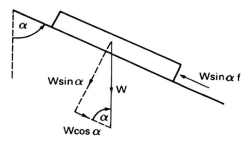

Fig. 11.7. Forces acting in a deviated well.

The same problem arises when running wireline tools (e.g. surveying instruments). The force distribution is shown in Fig. 11.7.

Downward movement will only occur if the axial component ($W \cos \alpha$) is greater than the drag ($W \sin \alpha f$). There will also be a frictional force resisting rotational movement of the drill string. To reduce the friction coefficient f and hence the torque and drag on the drill string, an oil-based mud, or a water-based mud with a significant percentage of lubricant added, is often used. Heavy-weight drill pipe is also used in horizontal wells because it is more flexible and less prone to sticking.

Numerous mechanical devices have been proposed to increase the driving force on the bit. A hydraulically operated drill collar has extended arms or anchors that grip the side of the borehole. A piston mechanism within the collar pushes down to advance the bit. A re-set mechanism allows the process to be repeated. It is not known whether such a system has been successfully tried on a horizontal well.

Controlling the well path
Most of the applications of horizontal wells already discussed in this chapter rely for their success on accurate placement within the reservoir. Precise directional control of the wellpath is therefore vital if the horizontal well is to be effective. Owing to the problems of exerting WOB and the tendency for the bit to be deflected by natural features such as bedding planes, it is not always possible to keep the well on course.

The 90° inclination is usually accomplished by having two or more build-up sections. These build-up sections can best be drilled with a downhole motor and bent sub. The bent subs generally have a larger offset angle (2–4°) than those used in conventional wells. Several different sizes of bent sub may be required to follow the planned trajectory.

Close monitoring of the well path using survey tools run on wireline would not be practical, because of:

(a) the time taken to run and retrieve the tools;
(b) the difficulty in lowering them down in a highly deviated hole.

MWD tools can provide both directional and logging data as the well is being drilled. A gamma-ray sensor is especially useful in detecting marker beds and for correlation with offset wells while the hole is being drilled.

Hole cleaning

As the inclination of the wellbore increases, so does the tendency for the drill cuttings to drop onto the Low Side of the hole. Continued build-up of cuttings will increase the risk of getting stuck pipe, since the drill collars will also tend to sag against the Low Side of the hole. A build up of cuttings will also cause problems when running logging tools or liners. The precautions taken to avoid stuck pipe in conventional wells apply also to horizontal wells. In addition to these, however, some other techniques may be employed.

Eccentric tool joints. As the drill string rotates, the eccentric tool joints will stir up any cuttings that have been deposited, returning them to the main flow stream.

Reverse circulation subs. These can be made up as part of the bottom hole assembly to divert flow from the drill string into the annulus to move cuttings off the side of the borehole.

Top drive systems. The pipe often becomes stuck when tripping out of the hole in unstable formations. This may also be due to cuttings settling out after circulation is stopped. When stuck pipe is detected, circulation and rotation of the string should begin as quickly as possible. A top drive system is capable of doing this much faster than making up the kelly and turning the rotary table. The power swivel can be quickly stabbed into the top joint of the drill pipe during a trip to allow rotation and circulation as the pipe is being pulled out of the hole. This process is known as "back-reaming" and it has been used successfully in many high-angle wells when tripping out of the hole. Top drive systems will be discussed more fully in the next chapter.

Mud properties and solids equipment. The properties of the drilling fluid must be carefully chosen to achieve good hole cleaning. The most important parameter is the yield point of the mud, which may have to be increased considerably in a highly deviated well. (A yield point/plastic viscosity ratio greater than 1.0 is commonly used.) Yield points of up to 30 lb/100 ft^2 have been reported. Turbulent flow in the annulus may also help to lift the cuttings from the Low Side of the hole.

It is essential that the cuttings lifted out of the annulus are effectively removed on surface before re-cycling. The solids content of the mud must be closely monitored. Efficient use of solids control equipment such as shakers, hydrocyclones and mud cleaners is essential.

Logging Problems

It has been observed that above an inclination of 50–60° logging tools run on wireline do not fall under their own weight. In order to run logs at higher inclinations, it is possible to pump the tools down through open-ended drill pipe or tubing. The drill string is run into the hole to the required depth. The logging tools are then lowered down inside the drill string with the aid of sinker bars. If the logging tools do not fall under gravity, pump pressure is applied to force them out the end of the drill pipe. This procedure is likely to damage the logging tools unless great care is taken. A more reliable system has been developed for logging high-angle wells (Fig. 11.8).

The logging tools are contained within a protective housing that is

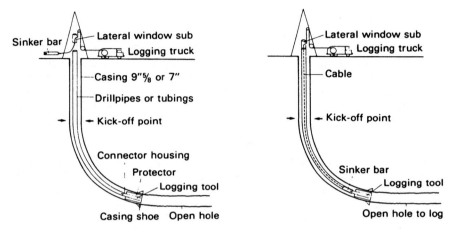

Step 1 = Running the logging tool with drillpipes

Step 2 = Bottom hole electrical connection between Logging tool and surface equipment

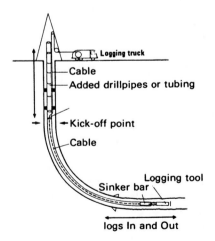

Step 3 = Running in (or out) by moving down (or up) the drillpipes

Fig. 11.8. Logging technique for horizontal wells (Simphor®, patented and registered trademark of the Institut Français du Pétrole).

mechanically fixed to the end of the drill string. On the upper end of the housing is an electrical (male) plug that will later connect with the female plug on the end of the wireline. The drill string is lowered to the top of the open hole section to be logged. At this stage a special side-entry sub is made up in the drill string to accommodate the conductor cable. A sinker bar and female connector are fitted to the end of the wireline that passes through the side-entry sub. The female connector is then pumped down the drill string until it latches with the male plug on the housing. The logging tools can then be powered up, ready to begin logging. More joints of pipe are then added to the string and the conductor cable passes through the annulus to the logging unit. The hole can be logged either running into the hole or pulling out of the hole. The advantage of this system is that standard logging tools can be used, and no special drill pipe equipment is necessary. This technique has been used successfully to log holes up to 90° inclination.

Completion Problems

The horizontal section of the well is usually completed by running a liner, which is tied back to the previous casing shoe set just above the reservoir. Even at modest inclinations, many operators have encountered problems in obtaining a good cement job on a liner. One of the major problems is to ensure good displacement of the drilling mud by the cement slurry. Before running the liner, therefore, the hole should be cleaned out and the mud conditioned. If mud solids, such as barite, are allowed to build up on the Low Side of the hole the cement slurry will by-pass this area leaving a mud channel within the cemented annulus. The cement slurry itself may be affected by gravity segregation, leaving free water on the upper side of the hole. This will allow communication of reservoir fluids along the annulus.

The displacement of mud by cement could be improved by reciprocating or rotating the liner. Since the setting tool is normally backed off prior to cementing it is not possible to reciprocate the liner. Furthermore, reciprocation of the pipe exerts piston forces along the hole that are undesirable and may cause the liner to stick. Rotation of the liner during cementing is possible by means of a special rotating liner hanger with sealed bearings. Correct use of centralizers will also improve the quality of the cement job.

Because of the problems associated with cementing the liner, some operators have chosen to run a pre-slotted liner. This greatly simplifies the completion operations, but it means that one particular zone cannot be isolated. This can prove to be a problem in a horizontal well, since more than one zone can be encountered owing to steeply dipping formations. External casing packers have been tried to seal off an unwanted zone (e.g. a vertical fracture producing water), but with little success. To be effective the pre-slotted liner must be carefully positioned away from any troublesome zore.

DRILLING PRACTICES FOR HORIZONTAL WELLS

Most of the horizontal wells drilled recently have employed basically the same techniques as used in conventional directional wells. Careful atten-

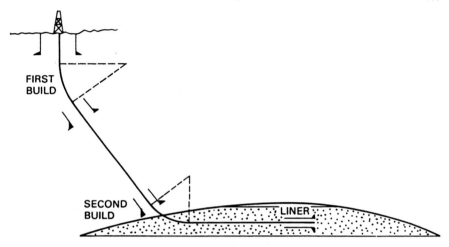

Fig. 11.9. Horizontal trajectory.

tion must be paid to the planning of the trajectory, directional surveying, use of downhole motors and turbines, selection of casing seats, choice of mud type and mud properties.

The well trajectory generally consists of more than one build-up section (Fig. 11.9). After setting the surface casing in the vertical part of the hole, a downhole motor and bent sub are used for the initial kick-off. A 2–3° bent sub will give sufficient build up rate (4–6° per 100 ft). Rotary building assemblies can also be employed in the build up section. An intermediate string of casing will then be set, and drilling will continue with a holding assembly. A second build-up section is drilled using a downhole motor and bent sub. This will bring the inclination to almost 90° at a point just above the reservoir. After setting another casing string the horizontal section can be drilled to the target.

Close directional control is required throughout, and so MWD tools are recommended. Any correction runs should be made by downhole motors. Turbodrills have been used to drill the straight sections of the hole. Rotary methods have also been used, although high torque may prove to be a problem.

A variety of different mud types have been tried with varying degrees of success. An oil-based mud reduces torque and drag and helps prevent differential sticking. However, some operators have discovered that a carefully formulated low-solids, water-based system (containing diesel or asphalt as a lubricant) can be equally effective. In some cases, the water based mud has been displaced to oil-based mud prior to drilling the horizontal section through the reservoir. The mud properties must be carefully controlled to provide good hydraulics and hole cleaning capability. Wellbore stability may also have to be considered where sloughing shales or unconsolidated formations are present. There is a risk of hole collapse if such formations are left unsupported by casing for long periods of time.

Drainhole Techniques

Drainhole drilling, or short-radius drilling, is another approach to horizontal drilling that requires a very sharp build-up section followed by a relatively short horizontal section. The radius of the build up section can be 20–40 ft (corresponding to a build-up rate of 2–3° per foot drilled). This rapid build up of angle is only possible because of a special bottom hole assembly.

Horizontal drainholes are generally drilled laterally from a vertical well. After logs or DSTs have been carried out to evaluate the formation, the well is plugged back to a point above the oil–water contact (Fig. 11.10). A special whipstock is then run into the hole on a tailpipe and oriented in the required kick-off direction. Alternatively, an orienting guide with a packer can be set. A kick-off assembly is then run into the hole, consisting of a bit, overgauge reamer, knuckle joint and articulated drill collars (Fig. 11.11). The knuckle joint introduces a bend in the assembly and more flexibility is provided by the special drill collars (referred to as "wiggly" collars). These are machined from standard drill collars by cutting a jig-saw pattern to create interlocking sections. The wiggly collars allow bending but still provide sufficient strength to transmit torque. The drilling fluid, however, must flow through a flexible hose that passes through the inside of the collars.

Once the assembly is run into the hole and deflected by the whipstock, it builds angle very quickly, reaching almost horizontal after drilling only 50–60 ft of hole. To avoid further angle-building, this assembly should be replaced by a stabilized BHA before reaching 90°. The reamer and knuckle joint are replaced in this second assembly by a full-gauge stabilizer. The wiggly collars are retained. This assembly is used to hold inclination through the reservoir section.

Normal wireline surveying techniques cannot be used because of the flexible hose that carries the drilling fluid. To take a survey in the drainhole

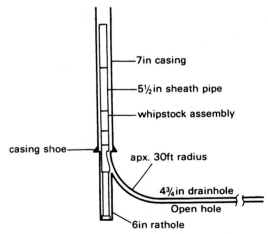

Fig. 11.10. Horizontal drainhole.

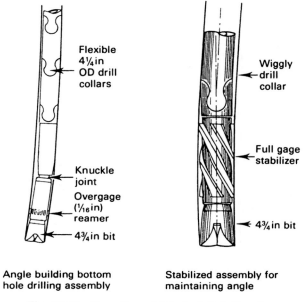

Fig. 11.11. Use of special "wiggly" drill collars.

section requires tripping out the bottom hole assembly. A single shot instrument can then be run on wireline with sufficient sinker bars to provide weight. The next BHA can be run in after the survey has been taken in the open hole section. Surveying is therefore a very time-consuming job in drainhole drilling. MWD tools currently available cannot be used under such conditions.

Recent horizontal drainholes have been up to 200 ft long. These small diameter holes ($4\frac{3}{4}''$) are generally left uncased. The same technique can be repeated further up the hole for multiple exploration and production from one wellbore. It is possible to drill out of cased hole after a window has been cut.

A summary of some recent horizontal wells drilled in Europe and North America is given in Table 11.1

QUESTIONS

11.1. A horizontal well is being planned to intersect a target whose horizontal displacement is 6000 ft at a TVD of 5500 ft. It is planned to have a 1200 ft horizontal section through the reservoir before reaching the target. The initial KOP will be at 1000 ft. There will be two build-up sections; the first at the KOP will be 3° per 100 ft and the second just above the reservoir will be 3° per 100 ft. Calculate the inclination over the tangent section and the total length of the well.

TABLE 11.1 Summary of Some Recent Horizontal Wells

Well	Location	Company	Year	Initial KOP (ft)	TVD of target (ft)	Hor. depth of target (ft)	MD of target (ft)	Length of hor. section (ft)	Max. build rate (deg/100 ft)
J-29X	Norman Wells, Canada	Esso	1979	98	1604	4616	5578	1860	6.1
K-142*	New Mexico, USA	Arco	1979	6037	6060	184	6145	106	300
Lacq 91	France	Elf	1981	279	2182	2707	4101	1214	3
Castera Lou	France	Elf	1983	5578	9515	4101	12025	886	1.2
HWP-2	Cold Lake, Canada	Esso	1985	131	1536	4922	5817	3334	4
Beck. 36	UK	BP	1985	328	3370	2812	5135	942	2.7
JX-2	Alaska, USA	Standard	1985	2800	8990	4878	11700	1363	2.5

* Horizontal drainhole.

11.2. List the main applications of horizontal drilling for offshore developments.

11.3. Explain some of the major factors that make a horizontal well more expensive than a vertical well.

11.4. Compare the drainage area of a platform whose wells have a maximum inclination of 55° with a platform whose wells have a maximum inclination of 70°. (Assume a Type I profile with a KOP of 1500 ft and a build up rate of 1.5° per 100 ft in each case and a target TVD of 10,000 ft.)

11.5. A horizontal drainhole must be drilled to locate a target 400 ft from the rig at a depth of 9500 ft (TVD). A build rate of 2° per ft can be achieved using a special bottom hole assembly. Calculate:
   (a) the depth of the kick off point;
   (b) the length of the horizontal section;
   (c) the total measured depth to the target.

11.6. Discuss the importance of drilling fluid properties in horizontal drilling.

FURTHER READING

"Mud and cement for horizontal wells', C. Zurdo, C. Georges and M. Martin, S.P.E. paper no. 15464.

"Horizontal drilling techniques at Prudhoe Bay, Alaska", J. P. Wilkerson, J. H. Smith, T. O. Stagg and D. A. Walters, S.P.E. paper no. 15372.

"Esso Resources horizontal hole project at Cold Lake", G. E. Bezaire and I. A. Markiw, 30th Annual Technical Meeting, Petroleum Society of CIM, Banff, May 1979.

"The use of horizontal drainholes in the Empire Abo unit", R. L. Stramp, S.P.E. paper no. 9221.

"Drainhole drilling", D. R. Holbert, Oil and Gas Journal, 9, 16 February 1981.

"The reservoir engineering aspects of horizontal drilling", F. M. Giger, L. H. Reiss and A. P. Jourdan, S.P.E. paper no. 13024.

"Problems associated with deviated wellbore cementing", S. R. Keller, R. J. Crook, R. C. Hant and D. S. Kulakofsky, S.P.E. paper no. 11979.

"Directional drilling technology will extend drilling reach", T. B. Dellinger, W. Gracely and G. C. Tolle, Oil and Gas Journal, 15 September 1980.

"In situ recovery from the Athabasca oil sands—past experience and future potential", D. A. Redford, Journal of Canadian Petroleum Technology, May–June 1985.

"Beckingham 36—horizontal well", P. Hardman, S.P.E. paper no. 15895.

"Torque and Drag in Directional Wells – Prediction and Measurement", Johansaik C. A., Freisen D. B., Dawson R., S.P.E. paper no. 11380 (1983).

"Uses and limitations of Drillstring Tension and Torque Models to Monitor Hole Condition", J. F. Brett, A. D. Beckett, C. A. Holt, S.P.E. paper no. 16664 (1987).

"Drillpipe buckling in Inclined Holes", R. Dawson, P. R. Paslay, S.P.E. paper no. 11167 (1982).

"Drilling the Coldlake Horizontal Well Pilot No. 2, R. R. MacDonald, S.P.E. Drilling Engineering, September 1987.

Chapter 12

CURRENT AND FUTURE DEVELOPMENTS

Directional drilling has benefited from many new tools and techniques over the past 10 years. In expensive offshore areas this new technology has been widely accepted as a means of improving efficiency and reducing drilling costs. In 1984 it was estimated that 25% of the world's gas and 20% of its oil came from offshore fields. As more and more of the easily accessible land-based reserves become depleted there will be a drift towards moving into more hostile environments and deeper offshore waters. Directional drilling is likely to play an increasingly important role in satisfying the world's demand for oil and natural gas over the next few decades.

Many of the research and development activities over recent years have flourished in a climate of high oil prices. The effect of lower oil prices and reduced drilling activity may impose some deceleration in the development of new tools. Even in the low-oil-price environment, however, there is the ever-present need to reduce drilling costs. In the short term, therefore, the focus is likely to be on making better use of the technology that already exists. Given a healthy economic environment, future research and development work will produce more advanced tools with better performance characteristics and greater potential for reducing costs.

DRILLING EQUIPMENT

Top Drive Units

Top drive systems are becoming more common on drilling rigs in order to save time and make it easier to drill highly deviated wells. A top drive system involves rotating the drill string using a power swivel instead of a kelly and rotary table. (Fig. 12.1). The power swivel is connected to the travelling block and both components run along a vertical guide track inside the derrick. Below the power swivel is a pipe-handling unit that

212

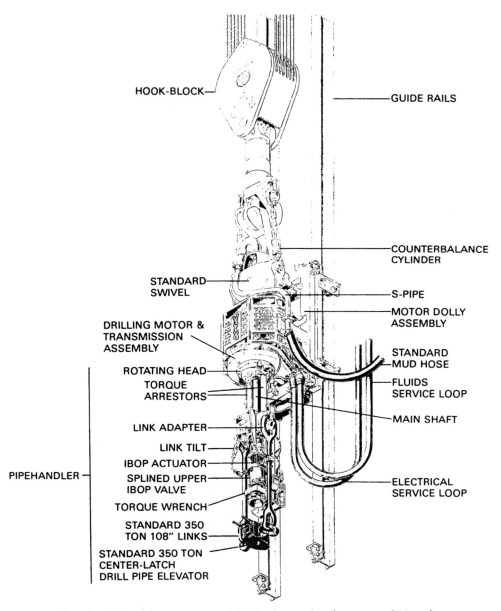

HOOK-BLOCK

GUIDE RAILS

COUNTERBALANCE CYLINDER

STANDARD SWIVEL

S-PIPE

MOTOR DOLLY ASSEMBLY

DRILLING MOTOR & TRANSMISSION ASSEMBLY

STANDARD MUD HOSE

ROTATING HEAD

FLUIDS SERVICE LOOP

TORQUE ARRESTORS

MAIN SHAFT

LINK ADAPTER

LINK TILT

IBOP ACTUATOR

PIPEHANDLER

SPLINED UPPER IBOP VALVE

ELECTRICAL SERVICE LOOP

TORQUE WRENCH

STANDARD 350 TON 108" LINKS

STANDARD 350 TON CENTER-LATCH DRILL PIPE ELEVATOR

Fig. 12.1. Top drive system used for back-reaming (courtesy of Varco).

consists of an elevator system and a hydraulic torque wrench to break out connections. The pipe handling unit allows the power sub to be stabbed into the drill string at any position in the derrick while tripping. A top drive unit enables the hole to be drilled in 90 ft stands, rather than 30 ft singles. When the power sub is made up to the top of a stand of pipe, the string is rotated from a point high up in the derrick. A connection is made after each 90 ft stand has been drilled. The procedure is as follows:

   (i) Set the slips and shut off the pumps.
  (ii) Break out the power sub from the stand using the pipe-handling unit.
 (iii) Unlatch the elevators, and raise block and power swivel up into derrick.
 (iv) Latch the elevators around the top of the next stand.
  (v) Stab the power sub into the top of the stand and make up the upper and lower connections.
 (vi) Pick up the string, pull the slips and commence drilling.

The advantages offered by a top drive unit are:

(a) By drilling with stands, two out of the normal three drill pipe connections are eliminated. This saves time and reduces wear on equipment.
(b) When tripping into the hole, any bridges or obstructions can be quickly reamed out by stabbing the power sub onto the drill string at any position in the derrick.
(c) Similarly, when pulling out of the hole in the event of stuck pipe, the power sub can be quickly connected to the string and circulation and rotation can begin immediately. It is therefore possible to back-ream through tight spots if necessary.

The capability of back-reaming makes a top drive unit attractive in high-angled wells, in which many stuck pipe and fishing problems could be avoided. The disadvantages of a top drive unit include:

(a) high initial cost of installation;
(b) added top side weight.

Top drive units have been installed on a number of North Sea platforms.

Downhole Motors and Turbines

The extensive use of PDMs and turbodrills seems likely to continue as directional wells become deeper and more highly deviated. Together with the benefits of less drill string wear and higher rotational speed at the bit, their use as deflection tools will become more important. There has been considerable interest recently in methods of steering downhole motors and turbines so that corrections can be made to the wellpath without tripping out the hole. This would clearly be advantageous in deep directional wells.

    Motor manufacturers are also aware of the need to improve reliability and performance. Better design of bearing assemblies has resulted in longer downhole operating times. Motors and turbines have been designed to operate in higher temperatures (e.g. geothermal wells). Multilobe, high torque motors are expected to find wider application.

The interaction between motors and bits will be more closely studied with a view to improving performance. New designs of PDC bits suitable for harder formations will also extend the applications of downhole motors.

Surveying Instruments

A wide variety of instruments already exists for surveying the borehole both during drilling and after casing has been set. Conventional single shots and multishots are becoming less common with the introduction of solid-state sensors (e.g. electronic multishots). The only instruments that can be used to survey the hole as it is being drilled are steering tools or MWD tools, both of which rely on magnetometers to measure azimuth. These are unreliable in the presence of casing, and so gyroscopic tools must be employed. One development that would find wide applications is a gyro-based instrument that could be placed behind the bit to monitor the wellpath continuously. The gyro would have to be designed to operate reliably despite the vibrations generated downhole while drilling. If the gyro could be incorporated in an MWD system, it would be a very powerful tool.

A variety of tools based on rate gyros or North-seeking gyros is used to survey the cased wellbore. Inertial navigation systems provide the most accurate method currently available, but the tool is restricted to large-diameter casing sizes. The problem here is to design the tool for a smaller-diameter hole, while retaining its accuracy. (The accuracy of a gyro reduces as its diameter reduces).

MEASUREMENT WHILE DRILLING

Although directional surveying is likely to remain the biggest application for MWD there is still some potential for growth in other areas. Multi-sensor tools are becoming more widely used.

MWD Sensors

Formation evaluation

One of the major deficiencies of MWD that has limited its wider application as a logging tool is the lack of a porosity sensor. A density or neutron device coupled with the existing gamma-ray and resistivity devices would provide a more complete suite of logs for interpretation of formation characteristics.

Another possible development is a caliper log that measures the diameter of the borehole. While drilling, this device could detect washouts and help in the design of hydraulics and cementing programmes.

Drilling safety

Another potential use of MWD that so far has received little attention is in improving detection of hazards. A number of possible sensors could be incorporated in the downhole tool.

(a) Influx detector: to detect the presence of formation fluids or gases within the wellbore (e.g. H_2S (hydrogen sulphide)).

(b) Pressure sensor: an accurate means of determining bottom hole pressure while drilling. This could also be useful in circulating out kicks.

(c) Pore pressure prediction: a sensor that could detect the presence of an overpressured zone before the bit actually penetrates the zone.

At present there is little commercial incentive to develop such sensors, but in exploration wells in new areas they could provide valuable information.

Another parameter that could help in directional control is the side force acting on the bit. By monitoring the side force with a downhole sensor, the effectiveness of a particular BHA could be measured. This information could also be incorporated in a computer model to analyse drill string behaviour and help in the selection of BHAs.

Downhole Tool Configuration

As more sensors become available, the need for a more flexible arrangement of downhole components becomes necessary. A modular system with replaceable parts would seem to be the most convenient arrangement. An operator could then choose the specific modules required for a particular job. The failure of one component should not affect the other components. A wireline retrieval system for replacing faulty components without having to trip out of the hole would also be a major benefit.

The reduction in size of the various components would make for easier rig-ups, and prevent space problems at the rig. There is also a need for a wider range of tools that can be run in smaller hole sizes (e.g. 6 in. and less in diameter). As holes become deeper, all MWD components will have to withstand higher temperatures and pressures.

The need for faster transmission of the data is a major problem facing all MWD companies. More efficient coding of the mud pulses is one approach currently under investigation. Other companies take the view that some rationalization of the data requirements is necessary. There may be no need for all the data being measured to be sent to surface in real time. Intermediate logging data, for example, can be measured while drilling, but stored in the tool's memory until retrieved when the tool is back on surface. For faster data rates, hybrid systems have been proposed that combine elements of mud pulse telemetry with hard-wire systems. New developments in MWD technology should not be allowed to reduce the overall reliability of the existing systems. Reliability will remain as one of the operator's prime requirements.

Data Integration and Application

The data supplied by a MWD system represents only one source of data from the drilling process. Other sources include surface measurements at the drill floor, mudlogging data and wireline logs (Fig. 12.2). All this information needs to be integrated into a database that is easily accessible both to the people on the rig and to company offices through satellite

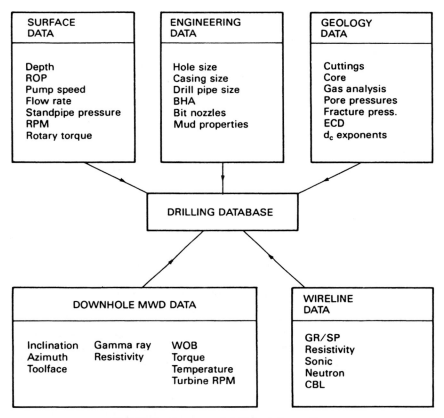

Fig. 12.2. Sources of drilling information.

communications. The data can then be used to make decisions as the well is being drilled.

One example of this is a method of determining bit wear from monitoring penetration rate and rotary speed at surface and combining this information with torque and WOB measurements downhole. A drilling model can be constructed for use in soft formations using milled tool bits which will detect bit wear. (Fig. 12.3). The effect of bit wear is easily disguised by relying only on surface measurements of torque, especially in directional holes. A problem such as a locked cone can easily go unnoticed until it is too late. The use of integrated MWD data, therefore, should allow the driller to realize what is happening at the bit and to make the correct decision as to when to pull out of the hole.

The use of a database can be extended further as an input for a drilling optimization programme that will select the drilling parameters to maximize efficiency. Surface measurements of torque and WOB are unreliable in directional wells, and could lead to erroneous results.

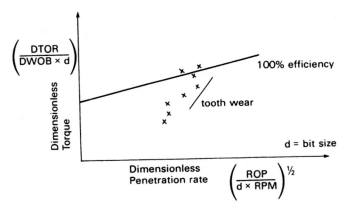

Fig. 12.3. Drilling model to measure tooth wear using both surface and MWD data.

A further development is directional optimization, in which the actual well path is compared with the planned well path. The amount of turn required to bring the well back on course could be calculated, along with the required toolface setting and drilling parameters. This information could then be sent to an adjustable bent sub in the bottom hole assembly which will automatically carry out the correction required. The idea of having a downhole tool that not only sends information to surface but also receives instructions from surface may lead to many future developments.

BHA ANALYSIS

The selection of BHA components is an important element in directional drilling. In most cases the choice is based on experience of drilling previous wells. It is usually the responsibility of the directional driller to select the number of stabilizers and their spacing within the BHA in order to achieve the desired trajectory.

Over recent years several attempts have been made to analyse BHA behaviour by developing mathematical models based on the physical properties of the components and the applied loading. Due to the large number of variables and the complex relationships involved, finite element techniques based on computer programs have been employed. This method allows a dynamic, three-dimensional analysis of the drill string by dividing the BHA into a large number of small discrete elements. By solving the force-displacement relationships for each element, taking into account boundary conditions, the forces acting along the entire drill string can be determined.

As with all methods of analysis, some assumptions and approximations have to be made. These are mainly related to the borehole environment (e.g. the friction coefficients between the borehole and the drill string have to be estimated). This type of analysis, however, has given reasonable

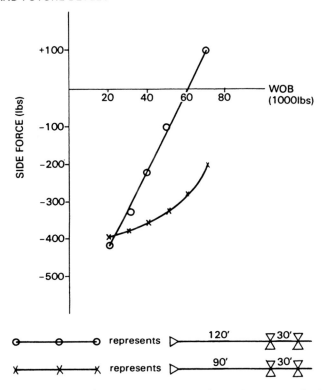

Fig. 12.4. BHA analysis results from finite element method.

results in comparison with actual field response. The major application of such a programme is to predict and assess the response of different BHA configurations to a given set of operating parameters. In Figure 12.4, for example, two different dropping assemblies are being compared in terms of the side force generated at the bit. Carrying out these calculations allows the optimum BHA to be selected in order to produce the required result. By improving the selection of BHAs, many hours of drilling time would be saved in avoiding extra trips and correction runs. This would also lead to reducing the wear on BHA components and improving the rate of penetration.

STEERABLE BHAs

The increased application of downhole motors and turbines as deflection tools run in combination with a bent sub has led to the concept of having an adjustable component within the BHA that is capable of altering the wellpath. This adjustable component is oriented in a specific way to change the azimuth or the inclination of the wellbore as required without having to pull out of the hole to change the BHA. Some developments are already underway.

Multi-angle Bent Sub

This special bent sub is made up in the BHA just above the downhole motor. It consists of two components that are connected by an articulated joint (Fig. 12.5). The lower sub is constructed so that it is able to rotate at an angle that is slightly offset from the vertial axis. Initially the tool is made up so that the upper and lower subs are aligned. When the lower sub is rotated it becomes locked in such a position that the two subs are offset by a small amount, thereby forming a bend in the BHA. Further rotation increases the size of the bend. There are ten possible positions for the lower sub to adopt, corresponding to angular offsets of 0° to 3° and back to 0° again.

The lower sub is actuated by a hydraulic device. The actuator creates a temporary surge in pressure that is transmitted to a shaft that turns the lower sub. At the end of the pressure surge, the lower sub is locked in that position until the next surge. The drill string must be lifted off bottom to change the angle of the bent sub; the flow rate is adjusted to provide the necessary increase in pressure to actuate the tool. Once in position, the mud pump is shut down and the lower sub is locked in position. It is therefore possible to operate the tool remotely from the rig floor.

This tool can be used to alter the amount of curvature in the build-up

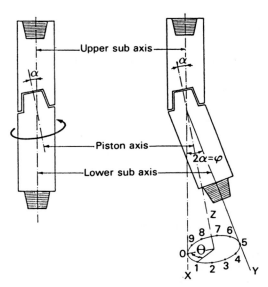

Fig. 12.5. Adjustable bent sub (Telepilote) (courtesy of IFP/SMFI Joint Development).

section of deviated wells without having to make a trip. It can then be returned to its original position for drilling the tangential section. It is therefore possible using this tool to drill to the KOP, build the required angle and start the tangential section all in one trip. Substantial savings are therefore possible in terms of the rig time that would otherwise be spent in enabling trips to change BHAs, or to replace a fixed-angle bent sub. Such a tool has applications in horizontal drilling, where more than one build up section may be necessary and frequent adjustments to the trajectory will have to be made.

Double-tilt Units (DTU)

When using a downhole motor as a deflecting tool, a bent sub can be placed on top of the motor, or a bent housing can be built into the lower end of the motor itself. This bent housing tilts the axis of the bit with respect to the axis of the rest of the drill string. This device can only be used for a steering run when the drill string is not being rotated.

By introducing a double-tilt unit (Fig. 12.6) immediately above the drive sub, the assembly can be used for both steering runs and conventional rotary drilling. For drilling a straight hole or tangential section the drill string is rotated 60–80 rpm from the rotary table, while the motor is also turning. The bend in the deflecting sub is cancelled out and it rotates concentrically, allowing the bit to drill ahead. When the bit has to be deflected during a kick-off or correction run, drill string rotation is stopped and the double tilt unit is oriented in the required direction. After completion of the steering run, the drill string is rotated and drilling ahead commences.

Offset Stabilizer

In the long tangential sections of directional wells, turbodrills are sometimes more cost-effective than conventional rotary drilling. When they are run with PDC bits, long bit runs and fast ROP can be achieved. Directional control over these long intervals is achieved by placing stabilizers at fairly close intervals along the length of the turbodrill. Variations in inclination or azimuth should be minimized during this interval. A turbodrill usually exhibits left-hand walk, which can be reduced by using shorter near-bit stabilizers. The left-hand walk can also be compensated by rotating the drill string. By carefully adjusting the stabilizers and applying drill string rotation it is possible, therefore, to steer the turbine in the required direction. This ability to make course corrections without changing the BHA, and thus avoid a trip, is obviously an advantage.

This led to the concept of the steerable turbodrill, which makes use of an offset stabilizer placed on the bearing assembly near the bit (Fig. 12.6). This stabilizer has three blades, two of which are full-gauge and placed 180° apart. The third blade is overgauge and is placed between the other two blades. When oriented in the required direction, the eccentric stabilizer will create a side force on the bit that will deflect the well path. The toolface of the offset stabilizer is oriented in the same way as a bent sub.

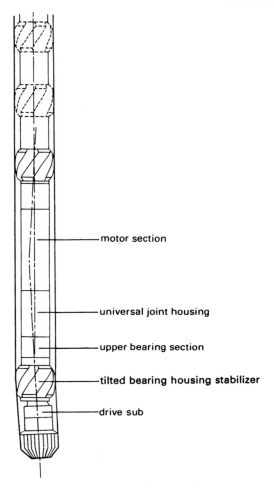

Fig. 12.6. Double tilt navigation sub (courtesy of Eastman-Christensen).

When drilling ahead, the drill string is rotated and the offset stabilizer has no effect on the wellpath. Where a correction run is necessary, the toolface is positioned using an MWD tool and the drill string is prevented from rotating. As the turbine drills, the bit will be deflected back onto the desired wellpath. After the correction run has been completed the drill string can be rotated as normal for drilling straight hole sections.

Intelligent Systems

Following on from these devices, the next step might be to have an interface between the adjustable component and the MWD system. A microprocessor within the downhole tool could be programmed to detect any divergence from the planned trajectory. An actuator within the tool

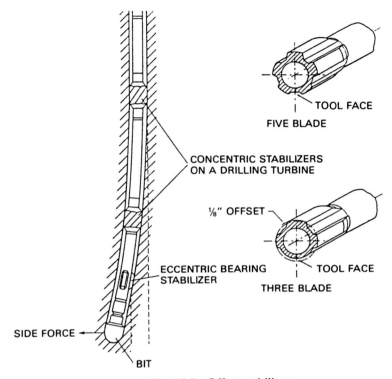

Fig. 12.7. Offset stabilizer.

could then operate the deflecting mechanism to bring the well back on course. The system should be flexible enough to operate with a conventional rotary assembly or with a downhole motor or turbine. Two-way communication with surface through some kind of telemetry link could also be built into the system (Fig. 12.8).

There are of course many practical problems to be overcome in implementing such a system. All the downhole components would have to be rugged enough to withstand the harsh drilling environment (vibration, temperature, pressure, etc.). A sophisticated control system would be required to supply all the relevant MWD information (azimuth, inclination, WOB, side force on bit, formation characteristics, etc.) to the microprocessor for analysis, and then actuate the deflecting device to bring about the required change in wellbore trajectory. The two-way communications link would provide a means of actuating the deflecting device if the control system failed. The system would also link up with surface actuators so that WOB and rpm can be adjusted.

The high cost of such a system would be offset by the considerable savings made by avoiding correction runs and the trips made to change-out bottom hole assemblies. Intelligent systems represent a major step forward in changing directional drilling from an art into a science.

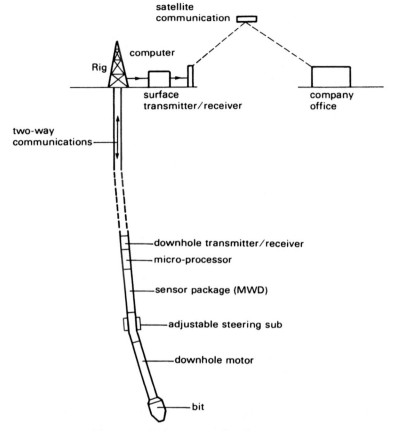

Fig. 12.8. Intelligent system for directional drilling.

FURTHER READING

"Today's achievements, and a look at future developments", C. C. Carson, *World Oil*, July 1984.

"ADMA uses new technology for turbodrilling", P. Juge, *Oil and Gas Journal*, 14 May 1979.

"New equipment, services reduce costs and improve efficiency", T. Muhleman, *World Oil*, April 1984.

"Bottom hole assembly analysis using the finite-element method", K. K. Millheim, S. Jord and C. J. Ritter, *JPT*, February, 1978.

"MWD Interpretation tracks bit wear", I. G. Falconer *et al.*, Oil and Gas Journal, 10 February, 1986.

"Some Technical and Economic Aspects of Stabiliser Placement", B. H. Walker, *JPT*, June 1973.

DEFINITIONS AND GLOSSARY OF TERMS

Abandon a well (v): to stop drilling or production operations when it becomes unprofitable. A wildcat may be abandoned after poor results from a well test. Cement plugs are placed in the wellbore to prevent fluid migration between different zones.

Abnormal pressure (n): a formation pressure that deviates from the normal formation fluid hydrostatic pressure. Such pressures may be classified as "subnormal" (lower than normal) or "overpressured" (higher than normal).

Accelerometer (n): a surveying instrument that measures components of the Earth's gravitational field.

Acidize (v): to apply acids to the walls of oil and gas wells to remove any material that might obstruct flow into the wellbore.

Adjustable choke (n): a choke in which a conical needle and seat vary the rate of flow.

Air drilling (n): a method of rotary drilling that uses compressed air as the circulating medium.

Angle unit (n): the component of a survey instrument used to measure inclination.

Annular preventer (n): a large BOP (q.v.) valve that forms a seal in the annular space between the wellbore and the drill pipe. It is usually installed above the ram type preventers in the BOP stack.

Annulus (n): the space between the drill string and the wellbore in open hole or cased hole.

Anticline (n): a configuration of folded and stratified rock layers in the shape of an arch. Often associated with a trap.

API (abbr): American Petroleum Institute. The leading standardizing organization on oilfield drilling and production equipment.

225

API gravity (n): a measure of the density of liquid petroleum products, expressed in degrees. It can be derived from the following equation:

$$\text{API gravity (degrees)} = \frac{141.5}{\text{specific gravity}} - 131.5$$

Azimuth (n): used in directional drilling as the direction of the wellbore measured in degrees (0–359) clockwise from True North or Magnetic North.

Back off (v): to disconnect a string of pipe by unscrewing one threaded piece from another.

Back pressure (n): the pressure maintained on equipment or systems through which a fluid flows.

Back up (v): (1) to hold one section of pipe while another is being screwed into or out of it (as in back up tongs); (2) a piece of equipment held in reserve in case another piece fails.

Badger bit (n): a specially designed bit with one large nozzle that can be used as a deflecting tool in soft formations (jet deflection bit).

Baffle plate (n): a replaceable metal ring inserted in the BHA (q.v.) to act as a landing collar for a survey tool.

Bail (n): a rounded steel bar that supports the swivel and connects it to the hook. May also apply to the steel bars that connect the elevators to the hook (links).

Ball up (v): to collect a mass of sticky material (drill cuttings) on components of drill string (especially bits and stabilizers).

Barge (n): a flat-decked, shallow draft vessel that might accommodate a drilling rig or be used to store equipment materials or for living quarters.

Barite (baryte) (n): barium sulphate ($BaSO_4$), a mineral used as a weighting material to increase mud weight (specific gravity = 4.2).

Barrel (n): a measure of volume for fluids. One barrel (bbl) = 42 US gallons = 0.15899 cubic metres.

Bearing (n): (1) a machine part that supports loads and allows other components to turn or slide; (2) used synonymously with azimuth or direction.

Bed (n): a geological term to specify one particular layer of rock.

Ball nipple (n): the uppermost component of the marine riser attached to the telescopic joint. The top of the nipple is expanded to guide drilling tools into the well.

Bentonite (n): a finely powdered clay material (mainly montmorillonite) that swells when mixed with water. Commonly used as a mud additive, and sometimes referred to as "gel".

Bent sub (n): a short piece of pipe whose axis is deviated 1–3° off vertical. Used in directional drilling as a deflecting tool. The scribe line on the inside of the bend defines the toolface.

BHA (abbr): bottom hole assembly.

Bit (n): the cutting element at the bottom of the drill string used for boring through the rock.

Bit breaker (n): a heavy metal plate that fits into the rotary table and holds the bit while it is being made up or broken out of the drill string.

Bit record (n): a report on each bit after it has been pulled out of the hole, containing details of its performance.

Bit sub (n): a short length of pipe installed immediately above the bit.

Bit walk (n): the tendency for the bit to wander off course by following the direction of rotation (usually to the right).

Blind rams (n): an integral part of the BOP (q.v.) stack, designed to close off the wellbore when the drill string is out of the hole.

Blocks (n): an assembly of pulleys on a common framework.

Blooey line (n): the discharge pipe from a well being drilled with compressed air. It carries the cuttings and reduces fire risk.

Blow-out (n): an uncontrolled flow of formation fluids into the atmosphere at surface.

BOP (abbr): blow-out preventer. One of several valves installed at the wellhead to control pressure in the event of a kick. A BOP has a full opening bore to allow drilling equipment and tools to pass through.

BOP stack (n): a unitized assembly of BOPs consisting of annular preventers and ram type preventers. For land drilling, the BOP stack is installed just below the rig floor; for floating rigs, the stack is positioned on the seabed.

Borehole (n): the hole made by the drill bit.

Bottom hole assembly (BHA) (n): the part of the drill string that is below the drill pipe. It usually consists of drill collars, stabilizers and various other components.

Bottom hole pressure (bhp) (n): the pressure at the bottom of the borehole, or at a point opposite the producing formation.

Box (n): the female section of a tool joint.

Brake (n): the device operated by the driller to stop the motion of the draw-works.

Break-out (v): to unscrew one section of pipe from another.

Bridge (n): an obstruction in the borehole usually caused by caving in of an unstable part of the wall.

BRT (abbr): below rotary table. Reference point for measuring depth.

Buck up (v): to tighten up a threaded connection.

Building assembly (n): a BHA (q.v.) specially designed to increase drift angle.

Build-up rate (n): the rate at which drift angle is increasing as the wellbore is being deviated from vertical. Usually measured in degrees per 100 ft drilled.

Build-up section (n): that part of the wellbore's trajectory where the drift angle is increasing.

Bumper sub (n): a drilling tool consisting of a short stroke slip joint that allows a more constant WOB when drilling from a floating rig.

By-pass (n): a route through which fluid can be pumped around a valve or other control mechanism (e.g. packer).

Cable-tool drilling (n): an earlier method of drilling used before the introduction of modern rotary methods. The bit was not rotated but reciprocated by means of a strong wire rope.

Caliper log (n): a tool run on wireline that measures the diameter of the wellbore. It may be used for detecting washouts, calculating cement volumes, or detecting internal corrosion of casing.

Cap rock (n): an impermeable layer of rock overlying an oil or gas reservoir and preventing the migration of fluids.

Cased hole (n): that part of the hole supported by a casing that has been run and cemented in place.

Casing (n): large-diameter steel pipe used to line the hole during drilling operations. Its purpose is to prevent the borehole from caving in, and to provide a means of producing the well.

Casing head (n): a heavy flanged connection that is installed on top of a casing string. Once it is installed, it provides a housing for the next casing string, which is suspended from it. Also called a spool.

Casing hanger (n): a special component made up on top of the casing string to suspend the casing when it is landed in the previous casing housing.

Casing shoe (n): a short section of steel filled with concrete and rounded at the bottom. This is installed on the bottom of the casing string to guide the casing past any ledges or irregularities in the borehole. Sometimes called a guide shoe.

Casing string (n): the entire length of all the casing joints run into the borehole. Several different sizes of casing string may be required.

Cathead (n): a spool-shaped attachment on a winch around which rope is wound. This can be used for hoisting operations on the rig floor.

Caving (n): collapse of the walls of the borehole. Also referred to as "sloughing".

Centralizer (n): a device secured around the casing at set intervals to centre the pipe in the hole.

Centrifugal pump (n): a pump consisting of impellor, shaft and casing, which discharges fluid by centrifugal force. Often used for mixing mud.

Centrifuge (n): an item of solids control equipment that separates out particles of varying density.

Cementing (n): the application of a liquid slurry of cement and water to various points inside or outside of the casing. Primary cementing is carried out immediately after the casing is run. Secondary cementing is carried out when remedial work is required.

Cement channelling (n): the creation of voids in the cement between the casing and the borehole, thereby reducing the effectiveness of the cement bond.

Cement head (n): a piece of equipment installed on the top of the casing that which allows the cement slurry to be pumped from the cement unit down the casing string. The cement head is also used for releasing the top and bottom cement plugs.

Cement plug (n): (**1**) a portion of cement placed at some point in the wellbore to seal off the well; (**2**) a device used during a primary cement job to separate the cement slurry from contaminating fluids in the casing. A wiper plug is pumped ahead of the slurry and a shut-off plug behind the slurry.

Chain tongs (n): a tool used by roughnecks on the rig floor to tighten or loosen a connection. The tool consists of a long handle and an adjustable chain that will fit a variety of pipe sizes.

Check valve (n): a valve that permits flow in one direction only.

Choke (n): an orifice installed in a line to restrict and control the flow rate.

Choke line (n): a pipe connected to the BOP (q.v.) stack that allows fluids to be circulated out of the annulus and through the choke manifold.

Choke manifold (n): an arrangement of pipes, valves and chokes that allows fluids to be circulated through a number of routes.

Christmas tree (n): an assembly of control valves and fittings intalled on top of the wellhead. The Christmas tree is installed after the well has been completed and is used to control the flow of oil and gas.

Circulate (v): to cycle drilling fluid through the drill string and wellbore, returning to the mud pits. This operation is carried out during drilling and is also used to condition the mud while drilling is suspended.

Clay (n): a term used to describe a material that is plastic when wet and has no well-developed parting along bedding planes. Such material is commonly encountered while drilling a well.

Clay minerals (n): the constituents of a clay that provide its plastic properties. These include kaolinite, illite, montmorillonite and vermiculite.

Closure (n): the shortest horizontal distance from a particular survey station back to the reference point.

Combination string (n): a casing string made up of various different grades or weights of casing (sometimes referred to as a tapered string).

Company man (n): an employee of an operating company whose job is to represent the operator's interests on the drilling rig (sometimes referred to as "drilling supervisor" or "company toolpusher").

Compass unit (n): the component of a survey instrument used to measure azimuth.

Completion (n): the activities and methods used to prepare a well for the production of oil or gas.

Conductor line (n): a small-diameter wireline that carries electric current. This is used for logging tools and steering tools.

Conductor pipe (n): a short string of casing of large diameter that is normally the first casing string to be run in the hole.

Connection (n): the joining of a section of drill pipe to the top of the drill string as drilling proceeds.

Continuous wave (n): one of the mud pulse telemetry systems used in MWD (q.v.).

Core (n): a cylindrical rock sample taken from the formation for geological analysis.

Core barrel (n): a special tool installed at the bottom of the drill string to cut and retain a core sample that is then recovered when the string is pulled out of the hole.

Correction run (n): a section of hole that must be directionally drilled to bring the well path back onto the planned course.

Course length (n): the measured depth between successive survey stations.

Crater (n): a large hole that develops at the surface of a wellbore, caused by the force of escaping gas, oil or water during a blow-out.

Crooked hole (n): a wellbore that has deviated from vertical, usually owing to geological factors.

Cross-over (n): a sub that is used to connect drill string components that have different threads or sizes.

Crown block (n): an assembly of sheaves or pulleys mounted on beams at the top of the derrick over which the drilling line is reeved.

Cuttings (n): the fragments of rock dislodged by the bit and carried back to surface by the drilling fluid.

Deadline (n): that part of the drilling line between the crown block and the deadline anchor. This line remains stationary as the travelling block is hoisted.

Deadline anchor (n): a device to which the deadline is attached, securely fastened to the derrick substructure.

Deflecting tool (n): a piece of drilling equipment that will change the angle and direction of the hole.

Degasser (n): a piece of equipment used to remove unwanted gas from the drilling mud.

Density (n): the mass of a substance per unit volume. Drilling fluid density is usually measured in kg/m^3, g/cc or ppg.

Departure (n): one of the coordinates used to plot the path of the well on the horizontal plane (along the x axis).

Derrick (n): a large load-bearing structure from which the drill string is suspended.

Derrickman (n): a member of the drilling crew whose workstation is on the monkey-board high up in the derrick. From there he handles the upper end of the stands of drillpipe being raised or lowered. He is also responsible for maintaining circulation equipment and carrying out routine checks on the mud.

De-sander (n): a centrifugal device for removing sand from the drilling mud.

De-silter (n): a centrifugal device for removing fine material (silt size) from the drilling mud.

Development well (n): a well drilled in a proven field to exploit known reserves. Usually one of several wells drilled from a central platform.

Deviation (n): a general term referring to the horizontal displacement of the well. May also be used to describe the change in drift angle from vertical (inclination).

Diamond bit (n): a bit with a steel body surfaced with diamonds to increase wear resistance.

Diesel (n): (1) a fuel used to power the prime movers on the rig; (2) an additive used in drilling mud and gunk material.

Differential sticking (n): one of the mechanisms causing stuck pipe by having excessive mud weight.

Dip (n): the angle at which the bed of the formation slopes away from the horizontal.

Direction (n): the orientation of the hole measured in the horizontal plane in degrees (0–90°) from North or South.

Directional drilling (n): the intentional deviation of a wellbore in order to reach an objective some distance from the rig.

Directional drilling supervisor (n): a specialist in directional techniques employed by an operator to ensure the target is reached.

Directional surveying (n): a method of measuring the inclination and direction of the wellbore by using a downhole instrument. The well must be surveyed at regular intervals to plot its course accurately.

Directional surveyor (n): surveyor hired by the operator to run survey tools in the hole.

Discovery well (n): the first well drilled in a new field that successfully indicates the presence of oil or gas reserves.

Displace (v): to move a liquid (e.g. cement slurry) from one position to another by pumping another fluid behind it.

Displacement fluid (n): the fluid used to force cement slurry or some other material into its intended position (e.g. drilling mud may be used to displace cement out of the casing and into the annulus).

Dog-house (n): a small enclosure on the rig floor used as an office by the driller and as a storage place for small items.

Dog-leg (n): a sharp bend in the wellbore that may cause problems tripping in and out of the hole.

Dog-leg severity (n): a parameter used to measure the amount of bending in the wellpath (usually given in degrees per 100 ft).

Dope (n): a lubricant for the threads of oilfield tubular goods.

Double (n): a section of drill pipe, casing or tubing consisting of two joints screwed together.

Downhole motor (n): a special tool mounted in the BHA (q.v.) to drive the bit without rotating the drill string from surface (see positive displacement motor).

Downhole telemetry (n): the process whereby signals are transmitted from a downhole sensor to a surface readout instrument. This can be done by a conductor line (as on steering tools) or by mud pulses (as in MWD tools (q.v.)).

Drag (n): The extra force required to move the drill string owing to the drill strings being in contact with the wall of the borehole.

Drag bit (n): a drilling bit that has no cones or bearings but consists of a single unit with a cutting structure and circulation passageways. The fishtail bit was an early example of a drag bit, but is no longer in common use. Diamond bits are also drag bits.

Drainholes (n): small-diameter horizontal wells drilled into a formation from a vertical borehole or shaft.

Drawworks (n): the large winch on the rig that spools off or takes in the drilling line and thereby raises or lowers the drill string.

Drift angle (n): the angle the wellbore makes with the vertical plane (see inclination).

Drill collar (n): a heavy, thick-walled steel tube that provides weight on the bit to improve penetration rates. A number of drill collars may be used beteen the bit and the drillpipe.

Driller (n): the employee of the drilling contractor who is in charge of the drilling rig and crew. His main duties are to operate the drilling equipment and direct rig-floor activities.

Drilling contractor (n): an individual or company that owns the drilling rig and employs the crew required to operate it.

Drilling crew (n): the men required to operate the drilling rig on one shift or tour. This normally comprises a driller, derrickman and two or three roughnecks.

Drilling fluid (n): the fluid circulated through the drill string and up the annulus back to surface under normal drilling operations. Usually referred to as mud.

Drilling line (n): the wire rope used to support the travelling block, swivel, kelly and drill string.

Drill pipe (n): a heavy seamless pipe used to rotate the bit and circulate the drilling fluid. Lengths of drill pipe 30 ft long are coupled together with tool joints to make the drill string.

Drill ship (n): a specially designed ship used to drill a well at an offshore location.

Drill stem (n): all the components from the swivel down to the bit.

Drill stem test (DST) (n): a test carried out on a well to determine whether or not oil or gas is present in commercial quantities. The downhole assembly consists of a packer, valves and a pressure-recording device, which are run on the bottom of the drill stem.

Drill string (n): the string of drill pipe with tool joints, which transmits rotation and circulation to the drill bit. Sometimes used to include both drill collars and drill pipe.

Drop-off section (n): that part of the well's trajectory where the drift angle is decreasing (i.e. returning to vertical).

Duplex pump (n): a reciprocating positive displacement pump having two pistons that are double-acting. Used as the circulating pump on some older drilling rigs.

Duster (n): a dry hole or non-productive well.

Dyna-drill (n): type of positive displacement motor (q.v.).

Dynamic positioning (n): a method by which a floating drilling rig or drill ship is kept on location. A control system of sensors and thrusters is required.

Easting (n): one of the coordinates used to plot the well's position on the horizontal plane (along the x axis).

Electric logging (n): the measurement of certain electrical characteristics of formations traversed by the borehole. Electric logs are run on conductor line to identify the type of formations, fluid content and other properties.

Elevators (n): latches that secure the drill pipe as it is being raised or lowered. The elevators are connected to the travelling block by links or bails.

Emulsion (n): a mixture in which one liquid (dispersed phase) is uniformly distributed in another liquid (continuous phase). Emulsifying agents may be added to stabilize the mixture.

Exploration well (n): a well drilled in an unproven area where no oil and gas production exists (sometimes called a "wildcat").

Fastline (n): the end of the drilling line that is attached to the drum of the draw-works.

Fault (n): a geological term that denotes a break in the subsurface strata. On one side of the fault line the strata has been displaced upwards, downwards or laterally relative to its original position.

Field (n): a geographical area in which oil or gas wells are producing from a continuous reservoir.

Filter cake (n): the layer of concentrated solids from the drilling mud that forms during natural filtration on the sides of the borehole. Sometimes called "wall cake" or "mud cake".

Filter press (n): a device used in the measurement of the mud's filtration properties.

Filtrate (n): a fluid that has passed through a filter. In drilling it usually refers to the liquid part of the mud that enters the formation.

Filtration (n): the process by which the liquid part of the drilling fluid is able to enter a permeable formation, leaving a deposit of mud solids on the borehole wall to form a filter cake.

FINDS (abbr): inertial navigation tool used in surveying.

Fish (n): any object accidentally left in the wellbore during drilling or workover operations, and which must be removed before work can proceed.

Fishing (n): the process by which a fish is removed from the wellbore. It may also be used for describing the recovery of certain pieces of downhole completion equipment when the well is being reconditioned during a workover.

Fishing tool (n): a specially designed tool that is attached to the drill string in order to recover equipment lost in the hole.

Flange up (v): to connect various components together (e.g. in wellheads or piping systems).

Flare (n): an open discharge of fluid or gas to the atmosphere. The flare is often ignited to dispose of unwanted gas around a completed well.

Flex joint (n): a component of the marine riser system that can accommodate some lateral movement when drilling from a floater.

Float collar (n): a special device inserted one or two joints above the bottom of a casing string. The float collar contains a check valve that permits fluid flow in a downward direction only. The collar thus prevents the back-flow of cement once it has been displaced.

Floater (n): general term used for a floating drilling rig.

Float shoe (n): a short cylindrical steel component that is attached to the bottom of a casing string. The float shoe has a check valve and functions in the same manner as the float collar. In addition, the float shoe has a rounded bottom that acts as a guide shoe for the casing.

Float sub (n): a check valve that prevents upward flow through the drill string.

Flocculation (n): the coagulation of solids in a drilling fluid produced by special additives or contaminants in the mud.

Fluid loss (n): the transfer of the liquid part of the mud to the pores of the formation. Loss of fluid (water plus soluble chemicals) from the mud to the formation can only occur when the permeability is sufficiently high. If the pores are large enough, the first effect is a "spurt loss", followed by the build-up of solids (filter cake) as filtration continues.

Formation (n): a bed or deposit composed throughout of substantially the same kind of rock and forming a lithologic unit.

Formation fluid (n): the gas, oil or water that exists in the pores of the formation.

Formation pressure (n): the pressure exerted by the formation fluids at a particular point in the formation. Sometimes called "reservoir pressure" or "pore pressure".

Formation testing (n): the measurement and gathering of data on a formation to determine its potential productivity.

Fracture (n): a break in the rock structure along a particular direction. Fractures may occur naturally or be induced by applying downhole pressure in order to increase permeability.

Fracture gradient (n): a measure of how the strength of the rock (i.e. its resistance to breakdown) varies with depth.

Fulcrum assembly (n): a bottom hole assembly designed to build angle.

Gas cap (n): the free gas phase that is sometimes found overlying an oil zone and occurs within the same formation as the oil.

Gas-cut mud (n): mud that has been contaminated by formation gas.

Gas show (n): the gas that is contained in mud returns, indicating the presence of a gas zone.

Gas injector (n): a well through which produced gas is forced back into the reservoir to maintain formation pressure and increase the recovery factor.

GCT (abbr): continuous guidance tool used in surveying.

Gel (n): a semi-solid, jelly-like state assumed by some colloidal dispersions at rest. When agitated, the gel converts to a fluid state.

Gel strength (n): a parameter used to measure the shear strength of the mud and its ability to hold solids in suspension. Bentonite and other colloidal clays are added to the mud to increase gel strength.

Geostatic pressure (n): the load exerted by a column of rock. Under normal conditions this pressure is approximately 1 psi/ft. This is also known as "lithostatic pressure" or "overburden pressure".

Guideline tensioner (n): a pneumatic or hydraulic device used to provide a constant tension on the wire ropes that run from the subsea guide base back to a floating drilling rig.

Guide shoe (n): a short cylindrical steel section placed at the bottom of a casing string. It is filled with concrete and rounded at the bottom to steer the casing past any irregularities in the borehole.

Gumbo (n): a naturally occurring clay that contaminates the mud as the hole is being drilled. The clay hydrates rapidly to form a thick plug that cannot pass through a marine riser or mud return line.

Gunk (n): a term used to describe a mixture of diesel oil, bentonite, and sometimes cement, that is used to combat lost circulation.

Gusher (n): an uncontrolled release of oil from the wellbore at surface.

Gyro multishot (n): a surveying device that provides a series of photographic discs showing the inclination and direction of the wellbore. It measures direction by means of a gyroscopic compass.

Gyro single shot (n): a surveying device that measures the inclination and direction of the borehole at one survey station. It measures direction by means of a gyroscopic compass.

Gyroscope (n): a wheel or disc mounted to spin rapidly about one axis, but free to rotate about one or both of the other two axes. The inertia of the wheel keeps the axis aligned with the reference direction (True North).

Heat shield (n): a protective barrel for survey instruments in high- temperature holes.

Hole opener (n): a special drilling tool that can enlarge an existing hole to the required diameter.

Hook (n): the large component attached to the travelling block from which the drill stem is suspended via the swivel.

Hopper (n): a large funnel-shaped device into which dry material (e.g. cement or powdered clay) can be poured. The purpose of the hopper is to mix the dry material with liquids injected at the bottom of the hopper.

Horizontal well (n): a directional well that reaches an inclination of 90°.

HWDP (abbr): heavy-weight drill pipe; thick-walled drill pipe used in directional drilling and placed between the drill collars and drill pipe.

Hydrostatic pressure (n): the load exerted by a column of fluid at rest. Hydrostatic pressure increases uniformly with the density and depth of the fluid.

Inclination (n): a measure of the angular deviation of the wellbore from vertical. Sometimes referred to as "drift angle".

Injection (n): usually refers to the process whereby gas, water or some other fluid is forced into the formation under pressure.

Impermeable (adj): preventing the passage of fluid through the pores of the rock.

Insert bit (n): a type of roller cone bit in which the cutting structure consists of specially designed tungsten carbide cutters set into the cones.

Intermediate casing (n): a string of casing set in the borehole to keep the hole from caving and to seal off troublesome formations.

Invert oil emulsion mud (n): a drilling fluid that contains up to 50% by volume of water, which is distributed as droplets in the continuous oil phase. Emulsifying agents and other additives are also present.

Isogonic line (n): an imaginary contour on the earth's surface joining places where the magnetic declination is the same.

Iron roughneck (n): a piece of rig-floor equipment that can be used to make connections and is automatically controlled.

Jack-up rig (n): an offshore drilling structure having extended legs that can be lowered to the seabed to support the hull and the drilling rig.

Jet bit (n): a drilling bit that has replaceable nozzles through which the mud is circulated. The high velocity of the mud leaving the nozzles improves drilling efficiency.

Jet deflection (n): a technique used in directional drilling to deviate the wellbore by washing away the formation in one particular direction. A special bit (badger bit) is used that has one enlarged nozzle that must be orientated towards the intended direction.

Jet sub (n): a tool used at the bottom of the drill pipe when the conductor pipe is being jetted into position (this method of running the conductor is only suitable where the surface formations can be washed away by the jetting action).

Joint (n): a single length of pipe that has threaded connections at each end.

Junk (n): debris lost in the hole, and which must be removed to allow normal operations to continue.

Junk sub (n): a tool designed to recover pieces of debris left in the hole.

Kelly (n): the heavy square or hexagonal steel pipe which runs through the rotary table and turns the drill string.

Kelly bushing (n): a device that fits into the rotary table through which the kelly passes. The rotation of the table is transmitted via the kelly bushing to the kelly itself. Sometimes called the "drive bushing".

Kelly cock (n): a valve installed between the kelly and the swivel. It is used to control a backflow of fluid up the drill string and isolate the swivel and hose from high pressure.

Kelly spinner (n): a pneumatically operated device mounted on top of the kelly which, when actuated, causes the kelly to rotate. It may be used to make connections by spinning up the kelly.

Keyseat (n): a channel or groove cut into the side of the borehole owing to the dragging action of the pipe against a sharp bend (or dog-leg).

Keyseat wiper (n): a tool made up in the drill string to ream out any keyseats that may have formed and thus prevent the pipe from becoming stuck.

Kick (n): an entry of formation fluids (oil, gas or water) into the wellbore, caused by the formation pressure exceeding the pressure exerted by the mud column.

Kill line (n): a high-pressure line connecting the mud pumps to the BOP (q.v.) stack, through which mud can be pumped to control a kick.

Killing a well (n): the process by which a well that is threatening to blow out is brought under control. It may also mean circulating water or mud into a completed well prior to workover operations.

KOP (abbr): kick-off point; The depth at which the wellbore is deliberately deviated from the vertical.

Latch (v): to attach the elevators to a section of pipe. Also used in connection with fishing jobs when the fishing tool grips the fish stuck in the hole.

Latitude (n): one of the coordinates used in plotting the wellpath on the horizontal plane (along the y axis).

LCM (abbr): lost circulation material.

Lead angle (n): the direction at which the directional driller aims the well to compensate for bit walk. Lead angle is measured in degrees left or right of the proposed direction.

Liner (n): (1) a string of casing that only extends back to inside the previous casing shoe and not back to surface as other casing strings do; (2) a replaceable sleeve that fits inside the cylinder of a mud pump.

Liner hanger (n): a slip type device that suspends the liner inside the previous casing shoe.

Location (n): the place at which a well is to be drilled.

Log (n): a systematic recording of data (e.g. driller's log, electric log, etc.)

Lost circulation (n): the loss of quantities of whole mud to a formation owing to caverns, fractures or highly permeable beds. Also referred to as "lost returns".

Magnetic declination (n): the angle between True North and Magnetic North. This varies with geographical location, and also changes slightly each year.

Magnetic multishot (n): a surveying instrument that provides a series of photographic discs showing the inclination and direction of the wellbore. It measures direction by means of a magnetic compass, and so direction is referenced to Magnetic North.

Magnetic North (n): the northerly direction in the Earth's magnetic field indicated by the needle of a magnetic compass.

Magnetometer (n): a surveying device that measures the intensity and direction of the earth's magnetic field.

Make up (v): to assemble and join components together to complete a unit (e.g. to make up a string of casing).

Make hole (v): to drill ahead.

Marine risen (n): the pipe that connects the subsea BOP (q.v.) stack with the floating drilling rig. The riser allows mud to be circulated back to surface, and provides guidance for tools being lowered into the wellbore.

Mast (n): a portable derrick capable of being erected as a unit, unlike a standard derrick which has to be built up.

Master bushing (n): an adaptor that fits into the rotary table and accommodates the slips and drives the kelly bushing.

Measured depth (MD) (n): the distance measured along the path of the wellbore (i.e. the length of the drill string).

Mill (n): a downhole tool with rough, sharp cutting surfaces for removing metal by grinding or cutting.

Milled-tooth bit (n): a roller cone bit whose cutting surface consists of a number of steel teeth projecting from the surface of the cones.

Monel (n): term used for a non-magnetic drill collar made from specially treated alloys so that it does not affect magnetic surveying instruments.

Monkey board (n): the platform on which the derrickman works when handling stands of pipe.

Moon pool (n): the central slot under the drilling floor on a floating rig.

Motion compensator (n): a hydraulic or pneumatic device usually installed between the travelling block and hook. Its function is to keep a more constant weight on the drillbit when drilling from a floating vessel. As the rig heaves up and down, a piston moves within the device to cancel out this vertical motion.

Mousehole (n): a small-diameter pipe or scabbard under the derrick floor in which a joint of drill pipe is temporarily stored for later connection to the drill string.

Mousehole connection (n): the procedure by which a new joint of pipe is added to the drill string. The new joint is already placed in the mousehole before the connection is made. The kelly is broken out from the top of the string and swung over to the mousehole where it is made up to the new joint. The kelly plus the new joint is then swung back to the rotary table where the new joint is made up to the top of the string and drilling can proceed.

MSL (abbr): mean sea level.

Mud (n): common term for drilling fluid.

Mud balance (n): a device used for measuring the density of mud or cement slurry. It consists of a cup and a graduated arm that carries a sliding weight and balances on a fulcrum.

Mud conditioning (n): the treatment and control of drilling fluid to ensure that it has the correct properties. This may include using additives, removing sand or other solids, adding water, and other measures. Conditioning may also involve circulating the mud prior to drilling ahead.

Mud engineer (n): usually an employee of a mud service company whose main responsibility on the rig is to test and maintain the mud properties specified by the operator.

Mud gun (n): a pipe that shoots a jet of drilling fluid under high pressure into the mud pit to agitate the mud or mix in additives.

Mudline (n): the seabed.

Mudlogging (n): the recording of information derived from the examination and analysis of drill cuttings. This also includes the detection of oil and gas. This work is usually done by a service company that supplies a portable laboratory on the rig.

Mud motor (n): a downhole component of the drill string that rotates the bit without having to turn the rotary table. The term is sometimes applied to both positive displacement motors and turbodrills (q.v.).

Mud pits (n): a series of open tanks through which the mud is cycled and conditioned. Modern rigs are provided with three or more pits, usually made of steel plate with built-in piping, valves and agitators.

Mud pump (n): a large reciprocating pump used to circulate the drilling fluid down the well. Both duplex and triplex pumps are used with replaceable liners. Mud pumps are also called "slush pumps".

Mud return line (n): a trough or pipe through which the mud being circulated up the annulus is transferred from the top of the wellbore to the shale shakers (q.v.). Sometimes called a "flowline".

Mud screen (n): shale shaker (q.v.).

Mule shoe (n): the guide shoe on the lower end of a survey tool; it locates into the key-way of the orienting sub. The survey tool can then be properly aligned with the bent sub.

MWD (abbr): measurement while drilling. A method of measuring parameters downhole and sending the results to surface without interrupting routine drilling operations. A special tool containing sensors, power supply and transmitter is installed as part of the BHA (q.v.). The information is transmitted to surface by a telemetry system using mud pulses.

Navi-drill (n): type of positive displacement motor.

Negative pulse (n): one of the mud pulse telemetry systems used in MWD (q.v.).

Nipple (n): a short length of tubing (generally less than 12 in.) with male threads at both ends.

Nipple up (v): to assemble the components of the BOP (q.v.) stack on the wellhead.

Non-magnetic drill collar (n): a collar that will isolate a magnetic compass from local magnetic effects (see monel).

Normal pressure (n): the formation pressure that is due to a normal deposition process in which the pore fluids are allowed to escape under compaction. The normal pressure gradient is usually taken as 0.465 psi per foot of depth from surface.

Northing (n): one of the coordinates used in plotting the position of the wellbore in the horizontal plane (along the y axis).

Nudge (v): to steer one well away from adjacent wells to avoid collisions.

Offshore drilling (n): drilling for oil or gas from a location that might be in an ocean, gulf, sea or lake. The drilling rig might be on a floating vessel (e.g. semi-submersible, drill ship) or mounted on a platform fixed to the seabed (e.g. jack up, steel jacket).

Oil-based mud (n): a drilling fluid that contains oil as its continuous phase with only a small amount of water dispersed as droplets.

Open hole (n): any wellbore or part of the wellbore that is not supported by casing.

Operator (n): the company that carries out an exploration or development programme on a particular area for which it holds a licence. The operator may hire a drilling contractor and various service companies to drill wells, and will provide a representative (company man) on the rig.

Orientation (n): the process by which a deflection tool is correctly positioned to achieve the intended direction and inclination of the wellbore.

Orienting sub (n): a special sub that contains a key or slot, which must be aligned with the scribe line of the bent sub. A surveying instrument can then be run into the sub, aligning itself with the key to give the orientation of the scribe line, which defines the toolface.

Ouija board (n): a device used to calculate the required orientation of a deflection tool.

Overburden (n): the layers of rock lying above a particular formation.

Overshot (n): a fishing tool that is attached to the drill pipe and is lowered over the fish and engages externally. An overshot can also be used to recover a wireline survey tool (single shot).

Packed hole assembly (n): a BHA (q.v.) that is designed to maintain inclination and direction of the wellbore.

Packer (n): a downhole tool, run on drill pipe, tubing or casing, which can be set hydraulically or mechanically against the wellbore. Packers are used extensively in drill stem tests, cement squeezes and completions.

Pay zone (n): the producing formation.

Pendulum assembly (n): a BHA (q.v.) that is designed to drop angle by allowing the drill collars to bend towards the low side of the hole.

Perforate (v): to pierce the casing wall and cement, allowing formation fluids to enter the wellbore and flow to surface. This is a critical stage in the completion of a well. Perforating may also be carried out during workover operations.

Perforating gun (n): a device fitted with shaped charges that is lowered on wireline to the required depth. When fired electrically from the surface, the charges shoot holes in the casing and the tool can then be retrieved.

Permeability (n): a measure of the fluid conductivity of a porous medium (i.e. the ability of fluid to flow through the interconnected pores of a rock). The units of permeability are darcies or millidarcies.

pH value (n): a parameter used to measure the acidity or alkalinity of a substance.

Pilot hole (n): a small-diameter hole that is later opened up to the required diameter. Sometimes used in directional drilling to control wellbore deviation during kick-off.

Pin (n): the male section of a threaded connection.

Pipe ram (n): a sealing device in a blow-out preventor that closes off the annulus around the drill pipe. The size of ram must fit the drill pipe being used.

Polycrystalline diamond compact bit (PDC bit) (n): a type of drag bit that uses small discs of man-made diamond as the cutting surface.

Pony collar (n): short drill collar, 10–15 ft. long.

POOH (abbr): pull out of hole.

Pore (n): an opening within a rock; is often filled with formation fluids.

Porosity (n): a parameter used to measure the pore space within a rock (usually given as a percentage).

Positive displacement motor (PDM) (n): a drilling tool that is located near the bit and used to rotate the bit without having to turn the entire drill string. A spiral rotor is forced to rotate within a rubber sleeved stator by pumping mud through the tool. Sometimes called a "Moineau pump" or "screw drill".

Positive pulse (n): one of the mud pulse telemetry systems used in MWD (q.v.).

Pressure gradient (n): the variation of pressure with depth. Commonly used under hydrostatic conditions (e.g. a hydrostatic column of salt water has a pressure gradient of 0.465 psi/ft).

Pressure relief valve (n): a valve that is preset to open when a certain limiting pressure is reached. Such valves are commonly installed on pipes, pumps, etc. Sometimes called safety valves or pop-off valves.

Primary cementing (n): placing cement around the casing immediately after it has been run into the hole.

Prime mover (n): an electric motor or internal combustion engine that is the source of power on the drilling rig.

Probe (n): a steering tool device that is lowered on the end of a conductor line to fit inside the orienting sub.

Production casing (n): the casing string through which the well is completed.

Propping agent (n): a granular material carried in suspension by the fracturing fluid and which helps to keep the cracks open in the formation after fracture treatment.

Protective casing (n): an intermediate string of casing that is run to case off any troublesome zones.

psi (abbr): pounds per square inch. Commonly used unit as a measure pressure.

Pup joint (n): a short section of pipe used to space out casing or tubing to reach the correct landing depths.

Rate gyro (n): a surveying tool that automatically aligns itself with True North (also called North-seeking gyro).

Rathole (n): (1) a hole in the rig floor 30–60 ft deep and lined with pipe. It is used for storing the kelly while tripping; (2) That part of the wellbore that is below the completion zone.

Reactive torque (n): the tendency of the drill string to turn in the opposite direction from that of the bit. This effect must be considered when setting the toolface in directional drilling when using a downhole motor.

Ream (v): to enlarge the wellbore by drilling it again with a special bit.

Reamer (n): a tool used in a BHA (q.v.) to stabilize the bit, remove dog-legs, or enlarge the hole size.

Reeve (v): to pass the drilling line through the sheaves of the travelling block and crown block and onto the hoisting drum.

Relief well (n): a directionally drilled well whose purpose is to intersect a well that is blowing out, thus enabling the blow-out to be controlled.

Reservoir (n): a subsurface porous permeable formation in which oil or gas is present.

Reverse circulate (v): to pump fluid down the annulus and up the drill stem or tubing back to surface.

Rig (n): the derrick, draw-works, rotary table and all associated equipment required to drill a well.

RIH (abbr): run in hole.

Riser tensioner (n): a pneumatic or hydraulic device used to provide a constant strain in the cables that support the marine riser.

RKB (abbr): referenced to kelly bushing. Term used to indicate the reference point for measuring depths.

Roller cone bit (n): a drilling bit with two or more cones mounted on bearings. The cutters consist of rows of steel teeth or tungsten carbide inserts. Also called a "rock bit".

ROP (abbr): rate of penetration, normally measured in feet drilled per hour.

Rope socket (n): the adaptor used to attach the survey tool to the wireline.

Rotary hose (n): a reinforced flexible tube that conducts drilling fluid from the standpipe to the swivel. Also called "kelly hose" or "mud hose".

Rotary table (n): the main component of the rotating machine that turns the drill string. It has a bevelled gear mechanism to create the rotation, and an opening into which bushings are fitted.

Roughneck (n): an employee of a drilling contractor who works on the drill floor under the direction of the driller.

Round trip (n): the process by which the entire drill string is pulled out of the hole and run back in again (usually to change the bit or BHA (q.v.)).

Roustabout (n): an employee of the drilling contractor who carries out general labouring work on the rig.

rpm (abbr): revolutions per minute; used as a measure of the speed at which the drill string is rotating.

Safety joint (n): a tool that is often run just above a fishing tool. If the fishing tool has gripped the fish but cannot pull it free, the safety joint will allow the string to disengage by turning it from surface.

Sail angle (n): a term used to describe the maximum drift angle in a directional well.

Salt dome (n): an anticlinal structure caused by an intrusion of rock salt into overlying sediments. This structure is often associated with traps for petroleum accumulations.

Sand (n): an abrasive material composed of small quartz grains. The particles range in size from $\frac{1}{16}$ mm to 2 mm. The term is also applied to sandstone.

Sandline (n): small-diameter wire on which light-weight tools can be lowered down the hole (e.g. surveying instruments).

Scratcher (n): a device fastened to the outside of the casing, and which removes mud cake and thus promotes a good cement job.

Scribe line (n): a reference line cut on the inside bend of a bent sub to indicate the toolface.

Semi-submersible (n): a floating drilling rig that has submerged hulls, but not resting on the seabed.

Shale (n): a finely-grained sedimentary rock composed of silt- and clay-sized particles.

Shale shaker (n): a series of trays with vibrating screens that allow the mud to pass through but retain the cuttings. The mesh must be chosen carefully to match the size of the solids in the mud.

Shear ram (n): the component of the BOP stack (q.v.) that cuts through the drill pipe and forms a seal across the top of the wellbore. These are used on floaters when the rig must be moved off location in an emergency.

Sheave (n): (pronounced "shiv") a grooved pulley.

Sidetrack (v): to drill around a permanent obstruction in the hole with some kind of deflecting tool.

Single (n): one joint of pipe.

Single shot (n): a survey tool that records the direction and inclination of the well at one point only.

Slips (n): wedge-shaped pieces of metal with a gripping element, used to suspend the drill string in the rotary table.

Slug (n): a heavy, viscous quantity of mud that is pumped into the drill string prior to pulling out. The slug will cause the level of fluid in the pipe to fall, thus eliminating the loss of mud on the rig floor when connections are broken.

Slurry (n): a pumpable mixture of cement and water. Once in position, the slurry hardens to provide an impermeable seal in the annulus, and supports the casing.

Spear (n): a fishing tool that engages internally and is used to recover stuck pipe.

Specific gravity (n): the ratio of the weight of a substance to the weight of the same volume of water.

Spider (n): a steel block with a tapered opening to accommodate the slips.

spm (abbr): strokes per minute; Rate of mud pump output.

Spool (n): a wellhead component that is used for suspending a string of casing. The spool also has side outlets for allowing access to the annulus between casing strings.

Spud (v): to commence drilling operations.

Squeeze-cementing (n): the process by which cement slurry is forced into place in order to carry out remedial work (e.g. shut off water-producing zones, repair casing leaks).

Stab (v): to guide the pin end of a pipe into the tool joint or coupling before making up the connection.

Stabbing board (n): a temporary platform erected in the derrick 20–40 ft above the drill floor. While running casing, one man stands on this board to guide the joints into the string suspended on the rig floor.

Stabilizer (n): a component placed in the BHA (q.v.) to control the deviation of the wellbore. One or more stabilizers may be used to achieve the intended wellpath.

Stage collar (n): a tool made up in the casing string, used in the second stage of a primary cement job. The collar has side ports that are opened by dropping a dart from surface. Cement can then be displaced from the casing into the annulus. Also called a "DV collar".

Stand (n): three joints of pipe connected together, usually racked in the derrick.

Standpipe (n): a heavy-walled pipe attached to one of the legs of the derrick. It conducts high-pressure mud from the pumps to the rotary hose.

Standpipe manifold (n): a series of lines, gauges and valves used for routing mud from the pumps to the standpipe.

Steering tool (n): surveying instrument used in conjunction with a mud motor to continuously monitor azimuth, inclination and toolface. These measurements are relayed to surface via conductor line, and shown on a rig-floor display.

Stimulation (n): a process undertaken to improve the productivity of a formation by fracturing or acidizing.

Stripping (n): movement of pipe through closed BOP's (q.v.).

Stuck pipe (n): drillpipe, collars, casing or tubing that cannot be pulled free from the wellbore.

Sub (n): a short, threaded piece of pipe used as a cross-over between pipes of different thread or size. Subs may also have special uses (e.g. bent subs, lifting subs, kelly saver sub).

Subsea wellhead (n): the equipment installed on the seabed for suspending casing strings when drilling from a floater (q.v.).

Suction pit (n): the mud pit from which mud is drawn into the mud pumps for circulating down the hole.

Surface casing (n): a string of casing set in a wellbore to case off any freshwater sands at shallow depths. Surface casing is run below the conductor pipe to depth of 1000–4000 ft (depending on particular requirements).

Surge pressures (n): excess pressure exerted against the formation owing to rapid movement of the drill string when tripping.

Survey (v): to measure the inclination and direction of the wellbore at a particular depth.

Survey depth (n): the measured depth at which the survey instrument is taking the measurements.

Survey station (n): the point at which a survey is taken.

Survey tool (n): the instrument that measures and records directional information.

Swabbing (n): a temporary lowering of the hydrostatic head owing to pulling pipe out of the hole.

Swedge (n): a connecting piece that joins pipes of different diameter.

Swivel (n): a component that is suspended from the hook. It allows mud to flow from the rotary hose through the swivel to the kelly while the drill string is rotating.

Syncline (n): a trough-shaped, folded structure of stratified rock.

Target (n): the objective, defined by the geologist, that the well must reach.

Target area (n): a specified zone around the target that the well must intersect.

Target bearing (n): the direction of the straight line passing through the target and the reference point on the rig. This is used as the reference direction for calculating vertical section.

TD (abbr): total depth.

Telescopic joint (n): a component installed at the top of the riser to accommodate vertical movement of the floating drilling rig.

Thread protectors (n): a device made of metal or plastic that is screwed on to pipe threads to prevent damage.

Tight formation (n): a formation that has low porosity and permeability.

Tongs (n): the large wrenches used to connect and disconnect sections of pipe. The tongs have jaws that grip the pipe, and torque is applied by pulling manually or mechanically using the cathead. Power tongs are pneumatically or hydraulically operated tools that spin the pipe.

Toolface (n): the part of the deflection tool that determines the direction in which deflection will take place. When using a bent sub, the toolface is defined by the scribe line. In jet deflecting, the toolface is the direction of the large nozzle on the bit.

Tool joint (n): a heavy coupling device welded onto the ends of drill pipe. Tool joints have coarse, tapered threads to withstand the strain of making and breaking connections and to provide a seal. They also have seating shoulders designed to suspend the weight of the drill string when the slips are set. On the lower end, the pin connection is stabbed into the box of the previous joint. Hardfacing is often applied in a band on the outside of the tool joint to resist abrasion.

Toolpusher (n): an employee of the drilling contractor who is responsible for the drilling and the crew. Also called rig superintendent.

Torque (n): the turning force applied to the drill string and causing it to rotate. Torque is usually measured in ft lb.

Tour (n): (pronounced "tower") an 8-hour or 12-hour shift worked by the drilling crew.

Trajectory (n): the path of the wellbore.

Trap (n): geological structure in which petroleum reserves may have accumulated.

Travelling block (n): an arrangement of pulleys through which the drilling line is reeved, thereby allowing the drill string to be raised or lowered.

Trip (v): to pull the drill string out of the hole, or to run it back in.

Trip gas (n): an accumulation of gas (usually a small amount) that enters the wellbore while making a trip.

Triplex pump (n): a reciprocating mud pump with three pistons that are single-acting.

True North (n): the direction of a line joining any point with the geographical North pole. Corresponds with an azimuth of 000°.

Tugger line (n): a small-diameter cable wound on an air-operated winch and which can be used to pick up small loads around the rig floor.

Turbodrill (n): a drilling tool located just above the bit and which rotates the bit without turning the drill string. The tool consists of a series of steel bladed rotors that are turned by the flow of drilling fluid through the tool.

TVD (abbr): true vertical depth. One of the coordinates used to plot the wellpath on the vertical plane.

Twist off (v): to sever the drill string by excessive force being applied at the rotary table.

UBHO sub (n): universal bottom hole orienting sub.

Underground blow-out (n): a situation arising when lost circulation and a kick occur simultaneously. Formation fluids are therefore able to enter the wellbore at the active zone and escape through an upper zone that has been broken down. (Sometimes called an "interal" blow-out.)

Under-ream (v): to enlarge the size of the wellbore below casing.

Union (n): a coupling device used to connect pipes.

Upset (n): the section at the ends of tubular goods where the OD is increased to give better strength.

Valve (n): a device used to control, or shut off completely, fluid flow along a pipe. Various types of valve are used in drilling equipment.

V door (n): an opening in one side of the derrick opposite the draw-works. This opening is used to bring in pipe and other equipment onto the drill floor.

Vertical section (n): the horizontal distance obtained by projecting the closure onto the target bearing. This is one of the coordinates used in plotting the wellpath on the vertical plane.

Viscometer (n): a device used to measure the viscosity of the drilling fluid.

Viscosity (n): a measure of a fluid's resistance to flow. The resistance is due to internal friction from the combined effects of cohesion and adhesion.

Vug (n): geological term for a cavity in a rock (especially limestone).

Washout (n): (1) wellbore enlargement due to solvent or erosion action of the drilling fluid; (2) a leak in the drillstring due to abrasive mud or mechanical failure.

Water back (v): to reduce the weight and solids content of the mud by adding water. This is usually carried out prior to mud treatment.

Water-based mud (n): a drilling fluid in which the continuous phase is water. Various additives will also be present.

Water injector (n): a well used to pump water back into the reservoir to promote better recovery of hydrocarbons.

Wear bushing (n): a piece of equipment installed in the wellhead that is designed to act as a bit guide and casing seat protector and to prevent damage to the casing hanger already in place. The wear bushing must be removed before the next casing string is run.

Weight indicator (n): an instrument mounted on the driller's console that gives both the weight on bit and the hook load.

Wellbore (n): a general term to describe both cased hole and open hole.

Wellhead (n): the equipment installed at the top of the wellbore from which casing and tubing strings are suspended.

Whipstock (n): a long, wedge-shaped pipe that uses an inclined plane to cause the bit to deflect away from its original position.

Wildcat (n): an exploration well drilled in an area where no oil or gas has been produced.

Wiper trip (n): the process by which the drill bit is pulled back inside the previous casing shoe and then run back to bottom. This may be necessary to improve the condition of the wellbore (e.g. smooth out any irregularities or dog-legs that could later cause stuck pipe).

Wireline (n): small-diameter steel wire that is used to run certain tools down into the wellbore; also called slick line. Logging tools and perforating guns require conductor line.

WOB (abbr): weight on bit; the load put on the bit by the drill collars to improve penetration rate.

WOC (abbr): waiting on cement. The time during which drilling operations are suspended to allow the cement to harden before drilling out the casing shoe.

WOW (abbr): waiting on weather. The time during which drilling operations must stop owing to rough weather conditions. Usually applied to offshore drilling.

Workover (n): the carrying out of maintenance and remedial work on the wellbore to increase production.

INDEX

Italic numbers refer to figures; bold numbers refer to tables.

ANSWERS TO NUMERICAL QUESTIONS

CHAPTER 3

3.1. Stiffness reduced from 5.91×10^9 in.^{2} − lb. to 5.87×10^9 in.^{2} − lb (0.8% reduction).

3.2. Section modulus of drill collar = 49.3 in.^{3}
Section modulus of drill pipe = 5.71 in.^{3}
ratio $= \dfrac{49.3}{5.71} = 8.6$ To reduce bending stresses use HWDP between drill pipe and collars.

3.3. Side force = −1006 lb

3.5. Change in azimuth = 4.9°
New inclination = 26.3°.

3.6. (a) Dog leg angle = 7.9°, course length = 158 ft.
(b) Final inclination = 14.4°.

3.7. Calculated toolface = 124° Right
Actual toolface = 94° Right (to account for reactive torque).

3.8. Toolface = 112° Right
Change in azimuth = 22°
Change in inclination = 0.6°

CHAPTER 4

4.1. (a) R = 3819.72 ft, x = −2.06°, y = 25.4°
inclination at end of build = 23.34°.
(b) Horizontal displacement = 312.57 ft
TVD = 1600 + 1513.32 = 3113.32 ft.
(c) Total measured depth = 1600 + 1556 + 8045 = 11,201 ft.

4.2. Vertical distance over tangent section can be calculated from

$$4000 - \frac{2R(1 - \cos 50°)}{\tan 50°} = 1639.03 \text{ ft}$$

Depth of KOP = 10,000 − 1000 − 1639.03 − (2 × 2194.56) = 2971.85 ft.

4.3. (a) R_1 = 1909.86 ft. R_2 = 2864.79 ft.
OQ = 1475.35 ft, OP = 8500 ft, QS = 4774.65 ft, PS = 7185.36 ft, x = 9.85°, y = 33.6°; inclination at end of build = 43.45°.
(b) At end of build TVD = 2813.45 ft, H = 523.35 ft.
At start of drop TVD = 8029.84 ft, H = 5464.88 ft.
(c) Total measured depth = 12,306.19 ft.

4.4. (a) Type I profile   max. inclination = 65.42°, MD = 11,730 ft.
   (b) Slant rig   max. inclination = 53.26°, MD = 10,750 ft.
4.5. R = 2291.83 ft
   Final inclination = 77.61°
   KOP depth = 3761.55 ft.

CHAPTER 5

5.1. (a) rotational speed = 471 rpm
   (b) torque output = 1375 ft lb
   (c) power output = 123 hp.
5.2. Pressure drop per stage $= \dfrac{252}{3} = 84$ psi.
5.5. Motor pressure drop = 300 psi
   Power output of 5/6 motor = 96 hp.

CHAPTER 6

6.1. (a) Power = 454 hp
   (b) Pressure drop = 1854 psi.
6.2. Speed = 689 rpm
   Torque = 4406 ft lb
   Pressure drop = 2277 psi
   Power = 577 hp.
6.4. Optimum speed = 750 rpm
   Maximum power = 128 hp.
6.5. Breakeven point = 230 ft.

CHAPTER 7

7.1. (a) S 55°W   or   235° magnetic
   (b) N 58°E   or   058° magnetic
7.2. (a) (i) in Gulf of Mexico, true bearing is S 62°W or 242°
      (ii) offshore Canada, true bearing is S 29°W or 209°
   (b) (i) in Gulf of Mexico, true bearing is N 65°E or 065°
      (ii) offshore Canada, true bearing is N 32°E or 032°
7.3. 30 ft monel required, with compass 3–4 ft below centre.
   If direction changes 2 monels required.
7.4. (a) 60 ft monel required
   (b) 90 ft monel required
7.6. Error = 27 ft, or 2.25 ft per 1000 ft.

CHAPTER 8

8.3. (a) Inclination = 1.16°
   (b) Azimuth = 59.6°, true azimuth = 52.6°
   (c) GTF = 40°L.
8.4. OTF = 136°

CHAPTER 9

9.1.
No.	N	E	TVD	VS
15	10.00	800.00	5900.00	796.08
16	10.58	833.32	5986.82	829.22
17	8.38	869.22	6072.58	865.19
18	−0.71	907.08	6157.04	903.68

9.2.
No.	Dog leg severity
16	2.6°/100 ft
17	2.6°/100 ft
18	3.5°/100 ft.

9.3.
	N	E	TVD	VS
(a) tangential	−3590.90	−590.90	5953.21	3130.46
(b) bal. tang.	−3573.26	−595.62	5958.52	3120.02
(c) min. curv.	−3573.48	−595.91	5959.00	3120.57

9.4. Closest distance = 10.8 ft.
9.5. (a) for a good gyro, lateral error = 15 ft
(b) for a poor gyro, lateral error = 85 ft.

CHAPTER 10

10.1. (a) 4.3° per 100 ft
(b) 2.5° bent sub required
(c) Allowing for reactive torque, toolface = 147°R.
10.2. 3°/100 ft
10.3. 5% in non-corrosive environment
55% in corrosive environment.
10.4. 1.2 % in non-corrosive environment
12.8% in corrosive environment.
10.5. 2618 lb.

CHAPTER 11

11.1. Inclination = 48.1°
Total measured depth = 9080.94 ft.
11.4. For 55° well, area = 9.7 square miles
For 70° well, area = 33.6 square miles.
11.5. (a) KOP depth = 9471.35 ft
(b) horizontal section = 371.35 ft
(c) total measured depth = 9887.7 ft.

CPSIA information can be obtained at www.ICGtesting.com
Printed in the USA
LVOW071045020613

336520LV00005B/39/A

9 780860 107163

DATE DUE

GAYLORD			PRINTED IN U.S.A.